AF332963

Designing Publics

Design Thinking, Design Theory
Ken Friedman and Erik Stolterman, editors

Design Things, A. Telier (Thomas Binder, Pelle Ehn, Giorgio De Michelis, Giulio Jacucci, Per Linde, and Ina Wagner), 2011

China's Design Revolution, Lorraine Justice, 2012

Adversarial Design, Carl DiSalvo, 2012

The Aesthetics of Imagination in Design, Mads Nygaard Folkmann, 2013

Linkography: Unfolding the Design Process, Gabriela Goldschmidt, 2014

Situated Design Methods, edited by Jesper Simonsen, Connie Svabo, Sara Malou Strandvad, Kristine Samson, Morten Hertzum, and Ole Erik Hansen, 2014

Taking [A]part: Experience-Centered Design and the Politics and Aesthetics of Participation, John McCarthy and Peter Wright, 2015

Design, When Everyone Designs: An Introduction to Design for Social Innovation, Ezio Manzini, 2015

Creating Frames: A New Design Practice for Driving Innovation, Kees Dorst, 2015

Designing Publics, Christopher A. Le Dantec, 2016

Overcrowded: Designing Meaningful Products in a World Awash with Ideas, Roberto Verganti, 2016

FireSigns: A Semiotic Theory for Graphic Design, Steven Skaggs, 2017

Designing Publics

Christopher A. Le Dantec

The MIT Press
Cambridge, Massachusetts
London, England

This book was set in Stone Serif and Stone Sans by Toppan Best-set Premedia Limited. Printed and bound in the United States of America.

Library of Congress Cataloging-in-Publication Data

Names: Le Dantec, Christopher A., author.
Title: Designing publics / Christopher A. Le Dantec.
Description: Cambridge, MA : The MIT Press, 2016. | Series: Design thinking, design theory | Includes bibliographical references and index.
Identifiers: LCCN 2016012609 | ISBN 9780262035163 (hardcover : alk. paper)
Subjects: LCSH: Human-machine systems—Design. | Industrial design. | Social ecology.
Classification: LCC TA167 .L34 2016 | DDC 620.8/2--dc23 LC record available at https://lccn.loc.gov/2016012609

10   9   8   7   6   5   4   3   2   1

To Renata

Contents

Series Foreword

As professions go, design is relatively young. The practice of design predates professions. In fact, the practice of design—making things to serve a useful goal, making tools—predates the human race. Making tools is one of the attributes that made us human in the first place.

Design, in the most generic sense of the word, began over 2.5 million years ago when *Homo habilis* manufactured the first tools. Human beings were designing well before we began to walk upright. Four hundred thousand years ago, we began to manufacture spears. By forty thousand years ago, we had moved up to specialized tools.

Urban design and architecture came along ten thousand years ago in Mesopotamia. Interior architecture and furniture design probably emerged with them. It was another five thousand years before graphic design and typography got their start in Sumeria with the development of cuneiform. After that, things picked up speed.

All goods and services are designed. The urge to design—to consider a situation, imagine a better situation, and act to create that improved situation—goes back to our prehuman ancestors. Making tools helped us to become what we are—design helped to make us human.

Today, the word "design" means many things. The common factor linking them is service, and designers are engaged in a service profession in which the results of their work meet human needs.

Design is first of all a process. The word "design" entered the English language in the 1500s as a verb, with the first written citation of the verb dated to the year 1548. *Merriam-Webster's Collegiate Dictionary* defines the verb "design" as "to conceive and plan out in the mind; to have as a specific purpose; to devise for a specific function or end." Related to these is the act of drawing, with an emphasis on the nature of the drawing as a plan or map, as well as "to draw plans for; to create, fashion, execute or construct according to plan."

Half a century later, the word began to be used as a noun, with the first cited use of the noun "design" occurring in 1588. *Merriam-Webster's* defines the noun as "a particular purpose held in view by an individual or group; deliberate, purposive planning; a mental project or scheme in which means to an end are laid down." Here, too, purpose and planning toward desired outcomes are central. Among these are "a preliminary sketch or outline showing the main features of something to be executed; an underlying scheme that governs functioning, developing or unfolding; a plan or protocol for carrying out or accomplishing something; the arrangement of elements or details in a product or work of art." Today, we design large, complex process, systems, and services, and we design organizations and structures to produce them. Design has changed considerably since our remote ancestors made the first stone tools.

At a highly abstract level, Herbert Simon's definition covers nearly all imaginable instances of design. To design, Simon writes, is to "[devise] courses of action aimed at changing existing situations into preferred ones" (Simon, *The Sciences of the Artificial*, 2nd ed., MIT Press, 1982, p. 129). Design, properly defined, is the entire process across the full range of domains required for any given outcome.

But the design process is always more than a general, abstract way of working. Design takes concrete form in the work of the service professions that meet human needs, a broad range of making and planning disciplines. These include industrial design, graphic design, textile design, furniture design, information design, process design, product design, interaction design, transportation design, educational design, systems design, urban design, design leadership, and design management, as well as architecture, engineering, information technology, and computer science.

These fields focus on different subjects and objects. They have distinct traditions, methods, and vocabularies, used and put into practice by distinct and often dissimilar professional groups. Although the traditions dividing these groups are distinct, common boundaries sometimes form a border. Where this happens, they serve as meeting points where common concerns build bridges. Today, ten challenges uniting the design professions form such a set of common concerns.

Three performance challenges, four substantive challenges, and three contextual challenges bind the design disciplines and professions together as a common field. The performance challenges arise because all design professions:

1. act on the physical world;
2. address human needs; and
3. generate the built environment.

In the past, these common attributes were not sufficient to transcend the boundaries of tradition. Today, objective changes in the larger world give rise to four substantive challenges that are driving convergence in design practice and research. These substantive challenges are:

1. increasingly ambiguous boundaries between artifacts, structure, and process;
2. increasingly large-scale social, economic, and industrial frames;
3. an increasingly complex environment of needs, requirements, and constraints; and
4. information content that often exceeds the value of physical substance.

These challenges require new frameworks of theory and research to address contemporary problem areas while solving specific cases and problems. In professional design practice, we often find that solving design problems requires interdisciplinary teams with a transdisciplinary focus. Fifty years ago, a sole practitioner and an assistant or two might have solved most design problems; today, we need groups of people with skills across several disciplines, and the additional skills that enable professionals to work with, listen to, and learn from each other as they solve problems.

Three contextual challenges define the nature of many design problems today. While many design problems function at a simpler level, these issues affect many of the major design problems that challenge us, and these challenges also affect simple design problems linked to complex social, mechanical, or technical systems. These issues are:

1. a complex environment in which many projects or products cross the boundaries of several organizations and stakeholder, producer, and user groups;
2. projects or products that must meet the expectations of many organizations, stakeholders, producers, and users; and

3. demands at every level of production, distribution, reception, and control.

These ten challenges require a qualitatively different approach to professional design practice than was the case in earlier times. Past environments were simpler. They made simpler demands. Individual experience and personal development were sufficient for depth and substance in professional practice. While experience and development are still necessary, they are no longer sufficient. Most of today's design challenges require analytic and synthetic planning skills that cannot be developed through practice alone.

Professional design practice today involves advanced knowledge. This knowledge is not solely a higher level of professional practice. It is also a qualitatively different form of professional practice that emerges in response to the demands of the information society and the knowledge economy to which it gives rise.

In a recent essay ("Why Design Education Must Change," *Core77*, November 26, 2010), Donald Norman challenges the premises and practices of the design profession. In the past, designers operated on the belief that talent and a willingness to jump into problems with both feet gave them an edge in solving problems. Norman writes:

In the early days of industrial design, the work was primarily focused upon physical products. Today, however, designers work on organizational structure and social problems, on interaction, service, and experience design. Many problems involve complex social and political issues. As a result, designers have become applied behavioral scientists, but they are woefully undereducated for the task. Designers often fail to understand the complexity of the issues and the depth of knowledge already known. They claim that fresh eyes can produce novel solutions, but then they wonder why these solutions are seldom implemented, or if implemented, why they fail. Fresh eyes can indeed produce insightful results, but the eyes must also be educated and knowledgeable. Designers often lack the requisite understanding. Design schools do not train students about these complex issues, about the interlocking complexities of human and social behavior, about the behavioral sciences, technology, and business. There is little or no training in science, the scientific method, and experimental design.

This is not industrial design in the sense of designing products, but industry-related design, design as thought and action for solving problems and imagining new futures. This new MIT Press series of books emphasizes strategic design to create value through innovative products and services, and it emphasizes design as service through rigorous creativity, critical inquiry, and an ethics of respectful design. This rests on a sense of understanding, empathy, and appreciation for people, for nature, and for the world we shape through design. Our goal as editors is to develop a series of vital conversations that help designers and researchers to serve business, industry, and the public sector for positive social and economic outcomes.

We will present books that bring a new sense of inquiry to design, helping to shape a more reflective and stable design discipline able to support a stronger profession grounded in empirical research, generative concepts, and the solid theory that gives rise to what W. Edwards Deming described as profound knowledge (Deming, *The New Economics for Industry, Government, Education*, MIT, Center for Advanced Engineering Study, 1993). For Deming, a physicist, engineer, and designer, profound knowledge comprised systems thinking and the understanding of processes embedded in systems; an understanding of variation and the tools we need to understand variation; a theory of knowledge; and a foundation in human psychology. This is the beginning of "deep design"—the union of deep practice with robust intellectual inquiry.

A series on design thinking and theory faces the same challenges that we face as a profession. On one level, design is a general human process that we use to understand and to shape our world. Nevertheless, we cannot address this process or the world in its general, abstract form. Rather, we meet the challenges of design in specific challenges, addressing problems or ideas in a situated context. The challenges we face as designers today are as diverse as the problems clients bring us. We are involved in design for economic anchors, economic continuity, and economic growth. We design for urban needs and rural needs, for social development and creative communities. We are involved with environmental sustainability and economic policy, agriculture competitive crafts for export, competitive products and brands for microenterprises, developing new products for bottom-of-pyramid markets and redeveloping old products for mature or wealthy markets. Within the framework of design, we are also challenged to design for extreme situations, for biotech, nanotech, and new materials, and design for social business, as well as conceptual challenges for worlds that do not yet exist such as the world beyond the Kurzweil singularity—and for new visions of the world that does exist.

The Design Thinking, Design Theory series from the MIT Press will explore these issues and more—meeting them, examining them, and helping designers to address them.

Join us in this journey.

Ken Friedman   Erik Stolterman
Editors, Design Thinking, Design Theory Series

Acknowledgments

This book would not have been possible without the help and support of many people, and unfortunately the thanks I offer here will be inadequate and incomplete. First I'd like to thank Keith Edwards, Beki Grinter, and Beth Mynatt for challenging and supporting me through the formative years of this work and for their continued friendship. Wendy Kellogg and her group including Mark Bailey, Jim Christensen, and Rob Farrell at IBM Research provided tremendous support; thank you all. My own thinking about publics and design has benefited tremendously from my conversations and friendship with Carl DiSalvo. Likewise, the students—present and former—in the Digital Media program at Georgia Tech, including Mariam Asad, Sarah Fox, Tom Jenkins, and Thomas Lodato, have each helped shape my thinking and have made the larger enterprise of academia so very worthwhile. And thanks to Mel Gregg for conversations on affect and for patiently reading through an earlier draft of the manuscript.

I am tremendously indebted to the shelter staff and residents who tolerated and trusted me during this project.

Finally, I'd like to thank my family: my parents François and Ellen and my sister Susan for a lifetime of encouragement; Renata, for her patience and support; Alexandr, Oliver, and Theo for reminding me to take a break and play a game or make pancakes once in a while. I could not have done this without them.

1 Introduction: Social Design in Public

In the summer of 2007, I spent several weeks beginning what would turn into a project of several years. The project began as a simple curiosity that arose from shifting practices enabled by personal computing—the selective attention and personal space afforded by things like iPods or the then revelatory iPhone. It struck me that in an urban environment these technologies were having an impact on people beyond urban professionals: people marching along, shielded by white earbuds, eyes fixed downward, face reflecting a blue glow in what is now the familiar signifier of having entered a non-place (Augé 1995).

Augé's notion of non-place resonated at the time because it provided an alternative account of the bubbles of intimate personal space these technologies created in public and transient locations. It is, in fact, the transience of the location mixed with the connection to personal encounters that sketches out the permanently liminal situation—moving through city streets or grocery stores or airports or bus stations while connected computationally to social worlds and interactions divorced from these mundane, everyday thoroughfares. And yet, while many of us are engrossed in these bubbles of personal curation, there are individuals for whom such locations are not transient and whose residence in them is more permanent and routine. The employee of an airport is not in transition; the airport is a destination and a stable and fixed location of her work. The clerk at the grocery store is not passing through and not able to remove himself by fiat of personal technology. Likewise, the urban homeless inhabit the public and liminal spaces of the city—its streets, its public squares, its transportation hubs—not in transit, but as a place where everyday life occurs (Tuan 1977; Spradley 1970). For all of these different occupations, the places they inhabit and labor within are simultaneously fixed places with immediate demands on attention, as well as non-places mediated by digital interactions, by data, and by easy escape to some other *here*. The technologies that enable that escape have forever changed the interpersonal interactions and obligations that play out in airports and grocery stores and other urban environments. Computing enables a society where the connected can divest themselves of place by rendering invisible individuals like the airport employee, the grocery store clerk, and the urban homeless person. The affluent have always had this capacity to selectively construct immediate social space—and it is certainly still the case that the wealthiest among us continue to posses the most choice in the matter—but computing has let some of that privilege leak into larger segments of society. Goods arrive at our doors in boxes disembodied from the social intercourse of shopping, we can hail transportation without a whistle or a wave, and we can arrange payment without the bother of negotiating tips. While many of the Internet-born services we use daily are welcome conveniences, they push those not participating further afield, out of the sight line. For the urban homeless in particular, this invisibility reinforces established social stratification and further marginalizes a variegated group who have been unable to marshal technology for their own purposes because of economic and social barriers.

Entangled in the move to personal technologies is the utopian ideal that our new digital arrangement will introduce an era of openness and transparency through the democratization of information, the enabling of new and meaningful social interactions, and digitally enabled modes of participation (Turner 2006). It would be foolish to argue that computing has not been beneficial for modern society, but the view is more complicated and the good it has produced is not unalloyed. Mobile phones and laptops arise out of a culture and economic environment built around consumption; they confer social significance as accoutrements of class membership, yet they also impact those without access and without the social and technical infrastructures that would enable access. For as much as these technologies enable important and tectonic shifts in a global ability to communicate, organize, and act in the world (Shirky 2008), they also stratify us based on our willingness or ability to incorporate them into our lives (Selwyn 2003).

It was from this assessment that I began my multiyear project to explore the design of technologies for the urban homeless. Granting that computing was already everywhere, there were a number of pragmatic questions regarding its impact on the urban homeless—how, in fact, did they use these devices and services? Were there obligatory interactions for employment? How did moves to push government and social services online impact both the homeless and social service providers? And, more fundamentally, what were the design considerations that demanded attention—how would such a project help reframe and motivate new ways of thinking about and theorizing the design of computational artifacts and services?

Theorizing Social Design

This last question is what drives this book. As I develop my response, it will be necessary to draw attention to some terms and grounding motivation that I will use in the following chapters. This book is, first and foremost, an effort to theorize design in what I call community contexts: those situations out in the world bounded by place, shared experience, and common cause. These might be long-standing, with a rich historical texture upon which to build design interventions, or they may be momentary and impermanent, lasting only as long as it takes an acute crisis to run its course. In any case, these settings are explicitly plural, and the constituencies and allies within them are aligned around particulars and should not be taken as given in the general sense. This is the primary practical difference between design in community settings and design in the commercial workplace, where the enterprise of work organizes people under a narrative of production that structures authority relations, incentives, and obligations.[1]

Carroll and Rosson (2007) point to the absence of a clear "us" and "them" as a defining marker of design in the community, as opposed to design in the commercial workplace. I would instead argue that it is a plural collection of "us" and "them" that make up participatory design in community settings, and that the aim of participatory design as working toward democratic workplaces is related but operationalized differently when placed in community settings (Ehn 1988). The kinds of concerns for quality of life and democratic participation have different implications in community settings than they do in industrial settings, and the causes and means for addressing shared concerns may be at a further remove from the individual when the categories of "management" and "labor" are not as obviously present as they are in the workplace (Dewey 1954; Ehn 1988).

Collectively, the alternative constraints of working in community settings and the characteristics of plurality and agonism come together in what I call a *public*. They form the basis for bounding an analytic and generative design space. This design space is inherently speculative as it seeks to imagine and enact alternative configurations around a set of shared conditions—see Chick and Micklethwaite (2011), Ehn, Nilsson, and Topgaard (2014), and Fuad-Luke (2009) for a variety of cases that cover sustainability, poverty, transportation, grass roots activism, and new forms civic organizing. As a focal point for design interventions, a public comes together and acts through design, resisting "accounts of what is [and] proposing that the condition under study could be transcended" (Rao, Krishnamurthy, and Kuoni 2014). This creates a trajectory in which designing publics attends to the cultural resources and mundane aspects of everyday life to move beyond modernism's pull of ready production (Hara 2014, 434; Papanek 1971, 20). It transpires both in the selective process of choosing where, how, and with whom to intervene (Löwgren and Reimer 2013, 165) as well as through the ongoing generative practices co-constituted within the setting that create new capacities to act.

Throughout this book, I work through the idea of publics via a case study on the design of a computational system. The value of publics—and of community-sited social design—is not limited, however, to how we think about design with respect to human-computer interaction. It is broader, and is technology-agnostic. The empirical case study presented here involves human-computer interaction simply because I am involved in human-computer interaction and it was the site where these ideas evolved. Computing is merely the material; the argument for how design circulates through communities to create new capacities to act collectively is an argument about the processes and commitments of design, not just the material objects of design (Fuad-Luke 2009, 153; Margolin and Margolin 2002; Morelli 2007).

That social design circulates within a larger discourse of design practice and aspiration is important. Starting with Papanek (1971), and tracing through more recent design scholarship (see for example Dunne and Raby 2013; Manzini 2015; Margolin and Margolin 2002; Morelli 2007), there has been a move to challenge the modernist vision of design and its roots in manufacture (of product and of desire for product). By creating alternative visions of design, we can bring into focus "other ways of managing our economic lives and the relationship among state, market, citizen, and consumer" (Dunne and Raby 2013, 9); moreover, such a focus instantiates "a design process intended to contribute to improving human well-being and livelihood" (Fuad-Luke 2009, 152). Arguably these endeavors are collective: they both require a diffusion of design as a common practice of problem setting *and* arise out of a time in which the material affordances of computing create new opportunities for participation in community, in culture, and in design (McCarthy and Wright 2015).

Publics, as I will develop more fully in the chapters to come, provide a vantage point from which to work through the theoretical grounds where design can intervene and operate as a form of collective action. Mine is an optimistic position that, in Manzini's terms, provides a way to mobilize design to create a "renewed pact between citizens and the state" (2015, 15), and drives at what Gilbert (2013) points toward in his considered reimagining of how collective forms of action (and governance) can be brought to bear against our contemporary social and political challenges. By focusing on forms of collective action through design, I am working through the ways design can be used to draw people together to contend with or resist shared social issues—to participate in the improvement of both their individual conditions and the conditions of their community (Morelli 2007). *Designing Publics* is both an analysis of how these publics come to be—the degree to which *they are designed* through intervention—and an account of the generative action publics take—the degree to which *they do design* as they mobilize and act in the world. There is an intentional and productive ambiguity captured in the title of this book: the book flips between an ethnographic lens examining how publics arise out of design intervention and a theoretical lens that focuses on how invention unfolds once a public is constituted. Taken together, this compound lens provides new resolution on how design and a diverse set of design practices can circulate outside sites of commercial production and into sites of social and collective action.

At its center, the argument is sociotechnical; it is about the social configurations that give rise to, and then evolve in response to, technological interventions—the *issues*, *attachments*, and *infrastructure* entangled when groups of people move to act. As a framework, publics provide a way to account for and involve actors at different scales (individuals, groups, and organizations), as well as the technologies and systems that bridge these scales in order to accomplish particular goals. Publics grant a perspective for understanding how actors, artifacts, and institutions gather around a unifying set of issues. This focus on issues, and on the action taken to reach a desired outcome with respect to those issues, is an important element of why publics are a potent frame for social design in community contexts: they help shift the focus from the entrenched power structures coincident with established stakeholders, and instead seek to highlight how diverse actors, including those holding opposing or contentious beliefs, come together to address those shared issues.

The focus on issues is both familiar to design and part of an ongoing migration of attention from working within stable social and technical settings toward embracing the wider set of dynamic relations that inform how issues arise, are interpreted, and are finally acted upon. This shift has been most evident within participatory design, in which the legacy of working within polarized environments—labor versus management, consumer versus producer—has been reconfigured in community-based settings where such bifurcation is no longer productive (Ehn 2008; Ehn, Nilsson, and Topgaard 2014). By turning to the concept of publics, the criteria for engagement and the definition of design spaces are moved away from alignment with one side in a pair of opposing structures, and instead position the design process to intervene in the articulation of *issues* and the effected network of diverse actors, artifacts, and institutions. When issues themselves become the primary concern, instead of actors as is typically the case (the individual consuming or using the designed artifact or system being the usual primary concern), design can more readily explore the *attachments* to those issues, the bonds between the actors, artifacts, and institutions effected by, and taking action on, a shared set of conditions. A further consequence of this shift of focus from entrenched stakeholders to issues and attachments is the move away from considering social design as primarily a method of product development and toward treating it instead as a political intervention in which design—the process—builds out social and technical capacities or *infrastructures*. Given this distinction, the object is not to design infrastructure (an artifact) but to engage in what Ehn (2008), citing Star and Ruhleder (1996), has referred to as infrastructuring, or the mechanism by which publics identify and subsume their sociotechnical resources: first articulating issues, then building out attachments, and finally integrating newly created resources for contending with the issues. The theoretical framing of publics is both the result and the means of social design; it creates a vocabulary for understanding how a community-sited design space comes into existence and changes over time; it provides pragmatic points of entry and intervention; it carries forward the capacities to act and intervene in the world.

When a public comes together around a shared set of issues—the externalities around which actors might coalesce—attachments are the internal mechanisms by which affiliations are created and resources identified, and infrastructuring is the work of building out future capacities for contending with issues built atop existing or newly created resources. These elements of a public—issues, attachments, and infrastructure—develop recursively, with partial definitions/articulations/solutions influencing and acting on each other, shifting the design space, and changing the constituency of the public. As the conception of the issues changes, as attachments are altered or rendered visible, and as different resources become embedded as infrastructure through acts of design, a public takes shape and becomes durable by way of these sociotechnical resources. As an organizing framework for social design, publics provide multiple points of entry for conceptualizing multiparty engagement by identifying and relating shared issues, supporting the articulation of attachments among diverse actors, and composing infrastructures that link issues and attachments (e.g., Asad and Le Dantec 2015; DiSalvo 2009; Le Dantec and DiSalvo 2013; Le Dantec and Fox 2015).

Threaded through the theoretical framing of publics are the set of practices rooted in community contexts that I refer to as *social design*. The term shares elements of social innovation, but I purposefully eschew the rhetoric of innovation, which valorizes the radical breakthrough over the steady, purposeful, and reflective practices of design. In calling out social design as a category or set of practices to attend to, I am making a selfish and pragmatic move to bound the kinds of design that might circulate in the service of community needs, simultaneously recognizing that the moral authority of design as a constructive intervention derives from a position of service to communities (Nelson and Stolterman 2002), and that as a mode and object of research, design is rarely composed of dramatic breakthroughs, but instead comprises many small moves that in aggregate produce new ways of acting in the world (Hara 2007; Norman and Verganti 2014).

The link between focusing on issue formation and social design is not a reductive act of problematizing and rationalizing the design space. Instead, it is a co-constructive act in which the relation between actors, artifacts, and institutions becomes a resource for scaffolding a public's engagement with a complex world. For example, to reduce homelessness to a set of serialized problems in need of solutions ignores the complexity, the institutionalized disadvantages, and the emotional consequences of poverty that confound—or simply make impossible—attempts by the homeless to implement dramatic changes to their lives. These issues are, as Povinelli puts them, the quasi events that accrete and derail long-term projects but are never individually in and of themselves momentous or catastrophic (Povinelli 2011, 144). Moreover, a reductionist problematization of homelessness delegitimizes many choices that individuals may make that, directly or not, play a role in their current situation, but are choices which, when removed, impinge on individual agency and sense of self-worth. Instead, social design as a set of practices focuses on issues, and turns to the community itself as designers able to confront present conditions, moving into a coproductive discourse that provides a social and cultural framing around how individual "problems" are enrolled in larger social and sociotechnical endeavors.

In a broader sense, the move toward developing social design—and returning to and integrating the foundations of Scandinavian participatory design when engaging in community settings—is a reaffirmation of the optimism of democratizing computing that was embedded in the early social movements behind the development of the personal computer (Turner 2006). However, democratization is a by-product neither of computation nor of a design process, but is instead a distinct goal of both the program of social design and of the resulting systems and artifacts (Björgvinsson, Ehn, and Hillgren 2010). It is something to strive for, seeking at each turn to build both social and technical tools and capacities that bridge participation from marginalized or disenfranchised individuals and groups. With this in mind, issue formation and the active voice of diverse stakeholders in that process become paramount.

Urban Homelessness and Social Services

For the case study around which the analysis in this book is built, the community carries both characteristics noted above: sited in a shelter for homeless single mothers and their children, the community had a long and rich history carried forward by the staff and volunteers who were the instrumental and emotional anchors; conversely, it was an ephemeral community, constantly changing as families moved through the shelter, arriving in crisis and transitioning through a series of programs and support structures to manage that crisis.

My choice to work at the shelter grew out of extensive fieldwork at a number of different service providers and was strategic as much as pragmatic: the shelter was well connected to a network of services in order to support the needs of the families in its care. The organization was an emergency shelter that provided thirty to ninety days of emergency housing for single women with their children. Up to eight families could be at the shelter at a given time, and the programs there focused on assisting the women in finding long-term housing (families were placed into transitional housing or apartments), employment, health care, child care, and assistance in securing any financial aid or social benefits for which they may have qualified.

There were a number of quirks about the rhythms of daily life at the shelter. The first, which later became an opportunity for design intervention, was that the shelter was closed from shortly after 8 in the morning until 5 in the evening. It was presumed that during this time the families were out meeting their responsibilities: mothers at work or actively seeking work, enrolling in support programs, or otherwise attending to the bureaucratic obligations of poverty, and children enrolled in school or child care. In the evenings, the families would arrive back at the shelter for dinner, followed by volunteer-led activities for the children (who ranged from infants up through high school) and a rotating program of group and individual counseling for the mothers as they worked through the particulars of their own circumstances. The schedule provided a structure that helped the families cope with crisis, it tried to preserve (or enforce) normalcy through work and school schedules, and it created a setting where the families could share, commiserate, and celebrate together.

The factors that contribute to homelessness—which is really just a symptom of poverty—are vast, systemic, and in many cases inherited (Beegle 2003). I would point out that my fieldwork at this shelter occurred between 2008 and 2011 and traced the housing collapse and subsequent recession that decimated the working class in the United States. By and large, these were the families resident at the shelter while I was there: single mothers or families separated due to economic conditions, having lost their job or struggling as underemployed household providers. Being homeless was a new experience for many of them, and there was a productive tension between the urgency to move past a moment of crisis, putting the shelter quickly in the rearview mirror, and a desire to linger so they might share newly acquired knowledge of how to cope and how to survive.

Social Issues and Social Design

New forms of computing—by which I refer broadly to the devices, applications, and platforms that have in many ways come to define how we live, work, and play—continue to rapidly change how we interact individually with one another and how we act collectively. On one hand, computing has enabled us to develop and recognize new forms of community that are divorced from traditional geographic and familial constraints (Bruckman 2006; Reingold 1993). On the other, it has helped existing communities— from close-knit social groups to individuals who merely cohabit public spaces (e.g., Grinter and Eldridge 2001; Paulos and Goodman 2004)—interact with each other in novel ways. These examples, from Grinter and Eldridge's study of teen texting habits, to Paulos and Goodman's notion of engaging "familiar strangers," to Reingold's account of early online communities, all share one common feature: access to, and through, computing.

Simply put, access to computing—the mobile phones, the Internet connections, the data—has opened new avenues of interaction and experience. It has created and populated a pervasive non-place where global political struggles are made visible while rooting that visibility in familiar experiences and language (Bennett and Segerberg 2012). Augé predicates his account of non-place on the observation that its inhabitants are transient; they are moving from one place to the next such that the definition of "here"—the context or site of use—is constantly in flux. Yet it is in this transition from one place to the next, from directly experiencing social conditions to having distant conditions brought near, that raises the possibility of new publics forming and acting through computing.

The notion of publics is potent precisely because it gives us a way to examine how people (and artifacts and institutions) are enlisted collectively through the interplay between a set of shared issues and the action taken to contend with those issues (Dewey 1954). Writing during a time of disillusionment with the democratic process, Dewey defined publics as groups whose needs were not being met by the present state of institutional authority or redress: it was a way to understand the formation of interest groups and to bound how those groups might come together to effectively operate within a democratic regime to effect desired changes. This notion is not only germane for describing community-state relations, it also provides a way to bound design, allowing for permeable boundaries around who is involved and a broad view of what is involved (Cohen 2005). It enables a pragmatic approach to establishing design programs in community settings, wedding social and technical approaches in the absence of a shared narrative or ideology. It makes explicit the relationship between a range of design practices and social issues—from those within human-computer interaction (McCarthy and Wright 2015), to design deployed to support activism (Fuad-Luke 2009), to the many configurations of social innovation (Ehn, Nilsson, and Topgaard 2014; Manzini 2015). Finally, it accounts for both social and technical actors, eschewing a reductive approach to problematizing a domain or context, instead providing avenues for issues and actors, artifacts and institutions, to evolve over time.

2 Publics and Their Issues, Attachments, and Infrastructures

The first question to answer in more detail is what, in the context of social design, is a public. There are a number of different ways scholars have applied the word *public* to groups of people, and a number of different relations implicated when a group of people is unified into a public. Here I would distinguish two broad ways in which other theorists have used the term *public* to understand and talk about various social settings. The first was born of queer theory, in which the words *public* and *counterpublic* are used to describe the multifaceted way audiences of a particular media artifact relate to each other, to the artifact, and to the author(s) of that artifact (Warner 2002). The second broad application of the word *public* concerns a body politic and ways of describing the relationship between a state or institutional entity and the community of citizens or individuals who are members of, or have dealings with, that entity (Lippmann 1993; Dewey 1954; Habermas 1991). It is this second set of definitions that I draw most directly upon for engaging in community-focused technology design; however, I also borrow freely from the ready building blocks to be found in Warner.

What Warner's framework of publics provides is a set of clear conditions that describes a kind of audience/performer relationship—that a public is self-organized, a relation between strangers, impersonal yet individually directed, constituted through attention, a social space bounded by direct and indirect contact with the media/content, and historically situated (Warner 2002). With respect to digital technologies, we can use this notion of publics as an alternative vocabulary for speaking about user experience in the context of shared systems so as to highlight the way different acts of sharing are performative expressions directed at particular and imagined audiences. Lindtner et al. (2011) do just this in their analysis of a system designed to support media-sharing practices within a dorm setting. Likewise, McCarthy and Wright's (2015) turn to publics and the participatory experiences of attending pop music concerts draws most directly from the framework Warner laid out. Similar analysis might fit readily into studies of social networking sites in which the audience ranges from known individuals (e.g., one's friends on Facebook) to a potentially broad set of unknown individuals who will come in contact with and potentially respond to posts and generated content (e.g., friends of friends on Facebook). This audience-performance relationship touches on a key element that is relevant to the definition of *public*: it emphasizes the porous boundary of a public, membership in which evolves through a relationship of production and consumption of shared content.

This fluid boundary is an important feature of publics: specifically, that a public's membership is not determined a priori, but instead evolves and arises out of a complex set of relations. For Lindtner et al. or McCarthy and Wright (and for Warner, on whom they build), content and media express those relations; for me it is the articulation and response to social issues that express them. My turn to social issues connects to the second broad conceptualization of publics, which has its roots in examining and describing the roles and responsibilities of public discourse and participation in governance and civil society. Tracing steps gradually backward, we can first examine the foundational notion of the public as set forth by Habermas, in which "the public" was defined as a discursive space (the "public sphere") within which concerned citizens took up issues of the common good through rational discourse (Habermas 1991). Critiques of Habermas's initial formulation of the public sphere have moved the discussion by highlighting the particular kinds of privileged social status that were necessary for a supposed rationally derived consensus (Calhoun 1993). As Fraser (1993, 113–114) points out:

[Through] the rhetoric of publicity and accessibility, the [original Habermasian formulation of the] public sphere rested, indeed was importantly constituted by, a number of significant exclusions. … [A] key axis of exclusion [was] gender … [in which] a new, austere style of public speech and behavior was promoted, a style deemed "rational," "virtuous," and "manly."

The broader point that Fraser and those she cites were making is that the early formulation of the public sphere rested on specific kinds of social relations—often those tied to "philanthropic, civic, professional, and cultural [societies that were] anything but accessible to everyone" (ibid., 114). These exclusions are important because they led Fraser, and later Warner, to employ the notion of "counterpublics" in order to make space for marginalized views and disenfranchised participants. While Fraser and Warner were working toward different ends, they shared a concern for recognizing diverse voices and pointing out that the boundaries of a public are far more permeable and uneven than the general public originally set out by Habermas.

In my use of the term *public*, I turn chiefly not to Habermas nor to Warner but instead to Dewey, who in 1927 set forth a view of participatory democracy that is instructive and relevant today in community settings where activism and local organization have become both sites of and resources for design. Dewey (1954, 15–16 and 35) sought to define a public not as a single common mass of people, but rather as a specific configuration of individuals bound by common cause in confronting shared issues:

The public consists of all those who are affected by the indirect consequences of transactions, to such an extent that it is deemed necessary to have those consequences systematically cared for. ... Those indirectly and seriously affected for good or for evil form a group distinctive enough to require recognition and a name. The name selected is The Public.

In Dewey's conception, then, publics are not a priori social groups. Rather, a public is a unique federation of people influenced or impressed upon by a specific set of conditions. As a public, they seek to address those conditions and their consequences. It is the combination of a set of shared conditions—issues—and actions taken to reach desired outcomes with respect to those issues that forms a public. With the focus on a plurality of publics, and the attention to social issues around which those publics form, we have the beginnings of a framework that embraces the contention, unevenness, and permeability that is ever present when accounting for the multiple voices, positions, and agendas present in any public discourse. In short, Dewey's pragmatic view is of a public that does not reductively treat social organization as a rational ideal, but instead seeks to work constructively within the messy and contentious reality of discourse where all voices—from mainstream to marginal—jockey to participate and arrive at desired outcomes.

While Dewey was concerned with publics as they related to statehood, it is useful to consider the formation of publics at scales that may only include particular communities. These communities may be physical (DiSalvo and Lukens 2009), they may form around virtual and distant interactions (DiSalvo and Vertesi 2007), or they may develop out of a set of shared conditions (Le Dantec et al. 2010; Le Dantec 2012). Encapsulated in the smaller scale of individual communities, we can still apply the general principles that arise out of groups of people identifying, articulating, and then acting to contend with a common set of issues. Much as Strauss's concept of "social worlds" provides a pragmatic framework for understanding the "real" social organization of large institutions (Strauss 1982; Strauss 1984), publics likewise help to identify and understand the conditions around which actors assemble themselves in the absence of a single overarching institutional organization. Moreover, this position expects and accounts for a messy amalgam of stakeholders, oppositional perspectives, and a need to constructively engage with contention and disagreement in the procedures around which such groups form. For the context of social design, this entails engaging with different actors such that the initial urge to reductively address acute symptoms is resisted in order to more broadly consider connections and less apparent opportunities for intervention.

Technology's role in fomenting a public occurs at the intersection of expressing issues and supporting action. Both in Dewey's pragmatist view of participatory democracy and in the early movements that gave rise to the social phenomena around the Internet, we find a deep optimism about society's ability to overcome challenges through sharing ideas (i.e., identifying issues) and engaging with each other (i.e., mobilizing action) (Dewey 1954; Turner 2006). However, sharing information and organizing a group for action are not easy. Even with the benefit of the widespread use of interactive technology, creating an identity for a public is a challenge:

Indirect, extensive, enduring and serious consequences of conjoint and interacting behavior call a public into existence [by] having a common interest in controlling these consequences. But the machine age has so enormously expanded, multiplied, intensified, and complicated the scope of the indirect consequences, [and] formed such immense and consolidated unions in action, on an impersonal rather than a community basis, that the resultant public cannot identify and distinguish itself. (Dewey 1954, 126)

The point being that, while publics can form around collective action, the myriad consequences facing contemporary society "produce both disaffection … and skepticism that collective action [can] solve pressing social problems" (Asen 2003, 178). This comes, in part, as Dewey observes, because "many consequences are felt rather than perceived; they are suffered, but they cannot be said to be known, for they are not, by those who experience them, referred to their origins" (Dewey 1954, 131).

Herein lies the balance to be struck when deploying design in the context of a public: as the ability to identify and express issues is made more accessible, support must also exist for connecting those affected by an issue to means of taking action to address that issue. As we work on projects of social design, we have an opportunity to provide tools that both amplify the ability to identify and articulate issues and empower action in response. This is, in short, a path to solving the problem of Dewey's public, of reconnecting the citizens with institutional entities; or more radically, of enabling citizens to wrest control over the resolution of issues from institutions that no longer act as effective intermediaries. While scholarship in public policy has often been unsuccessful in resolving Dewey's problem (Fung 2006, 17), there is an opportunity to seek solutions that bring design—and the design of new technologies or new relations to technology—to bear on the challenge of reconnecting state and society in a pragmatic and participatory way.

Sackman (1968) made this very point when he suggested that real-time computing could be the tipping point for supporting and instigating public action. Asserted in light of command and control systems of the mid-1960s, Sackman's vision of the critical role of computing in shaping public action is still relevant (and unfolding): publics can be constituted and supported with technologies that enable access to information, provide means of distributed information production, and include social mechanisms to identify and sustain individual members and help mobilize and organize others around their common issues. It is this idea that has brought publics to the fore as interest in social movements and political action has grown in contemporary design (e.g., Binder et al. 2011; DiSalvo 2012; Ehn, Nilsson, and Topgaard 2014; Fuad-Luke 2009; Löwgren and Reimer 2013; Manzini 2015). This is, undoubtedly, an optimistic position to hold, and it is important to point out very explicitly that the optimism does not come from the mere presence or application of computing, but from the careful process of considering a design space that is both social *and* technical. The inventive responses to issues that might come through social design, through the application of computing to present and new forms of advocacy and activism, through the diffusion of alternative forms of democratic participation, and through the inevitable (but by no means uncritically accepted) disruption of existing institutions are not born of a deference to computing as inherently democratizing, but from examining the whole of the sociotechnical landscape as a rich space for design invention.

The relevance of publics to design lies in the calibration of design to dynamic constituencies—regardless of the material practices that might be deployed in a given context (e.g., computing, public policy, architecture, etc.). The concern here is not the use of computing as such, but creating a fluid entity that forms and unforms in response to the articulation of evolving social conditions, with multiple points of entry for working in community settings. Like Warner's publics that emerge through the production and consumption of various media (and mediated) performances, and extending out of Dewey's publics of civic engagement and democratic participations, publics in design arise and respond to social conditions that enlist and are enlisted by the joint participation of artifacts, people, and institutions.

The Primacy of Issues

The engine behind the dynamic and contingent nature of publics is the presence and evolution of issues. This fluidity is realized by a dynamic cast of actors who are enlisted in order to contend with the effects of particular issues, and these issues are what make publics a useful perspective for social design by providing a stable theoretical frame around a dynamic context. Issues, then, are a base element needed to constitute a public, and it is the process of articulating and relating to a set of common social issues that begins to bring a given public into relief.

We can find one example of how issues drive publics in the development of genetically modified organisms for agriculture and food production. The issue, even before GMOs were put into widespread use in food production, was enough to catalyze the formation of a public that confronted the anticipated ecological, economic, and ethical consequences (Dryzek 2009). The fact that this action took place around objects that had yet to exist—or yet to exist on a mass scale—highlights the power of issues as an organizing principle; issues need not only be about present conditions, but may also be about perceived future conditions that have not yet come to pass (Dryzek 2009; Latour 2007; Latour and Weibel 2005).

This focus on the future is also present in the context at the heart of this book: much of the work done in providing social services is about planning and providing resources to head off the future needs of those seeking service. After an initial effort to establish some sort of stability, caseworkers and clients work to identify the issues that need to be resolved (typically a mix of education/job training, financial stability, and ongoing counseling for individuals or families). The issues identified are in part exigent concerns (e.g., finding housing, becoming employed, securing health care, enrolling children in school), but they extend to planning for how to maintain the temporary, scaffolded stability found in the shelter once the families return to living on their own and became more removed from the immediate care of support networks. As different groups of individuals become involved, the shape of the issues changes (Le Dantec et al. 2010; Le Dantec and Edwards 2008a; Le Dantec and Edwards 2008b): employment counselors focus on the role of employment as a primary issue, family counselors look at family relations and the formation of strategies to help these families care for their children, caseworkers take a broader view as they help the clients navigate multiple programs of service and care as best they are able, volunteers respond to immediate material needs such as food and clothing, and the homeless families are consumed by reacting to these issues as immediate crises. In fact, much of the counseling provided is meant to help the affected families rationalize the issues with which they are confronted so that as they work through the initial emotional response to crises, they can begin to plot a path that can be sustained over a longer period. In this way, issue definition is at the heart of homeless intervention activities and becomes a natural point of entry for social design—more than simply looking to support one set of stakeholders over another.

Whether within a program of design or not, issue formation shapes the discourse of a public by creating trajectories for potential solutions. As Marres (2007, 761) points out, "issue definition [is] the decisive factor in democratic institutional politics, as it determines which actors can get involved in political process, and on what terms." So, as an increasingly digital world removes the need to cling to political territories, we arrive at a position in which alliances and allegiances are more productively issue-based. No matter the geographic or political scope of those involved in the work of issue formation, power and authority dynamics emerge through issue formation as groups of individual or collective interests become involved. This underscores the importance of looking at how those issues are formulated and expressed: who was involved, what kinds of processes (formal or otherwise) were used, and what results were considered. Callon, Lascoumes, and Barthe's (2009) notion of "hybrid forums" addresses some of these questions, looking specifically at the articulation of issues amid heterogeneous actors. These agglomerations of diverse actors lead to a coincidence of different ways of knowing about and acting in the world— be it as lay community member, technical expert, or political figure. The intersection of these different, and at times incompatible, ontologies first shapes the definition of the issues and then the trajectory of action, invention, and development taken to address those issues. The relations between actors engaged in issue formation then shape, much more definitively, the potential responses to the issue, such that the main work of discourse lies in issue formation rather than in negotiating solutions. This dynamic is taken up thoroughly by Marres (2007), who traces out the importance of issue formation as the foundational activity of discourse as viewed by scholars in science and technology studies and political science.

As issue formation is of utmost importance in shaping community-based design projects, we need to explicitly acknowledge and work with the contentious and political dynamics that arise within communities. This form of productively confronting diverse views, of transforming antagonism into agonism in which the polyphony of views, values, and ways of knowing are turned into productive spaces for social design, fits within early models of participatory design as it took shape in the changing industrial settings of Scandinavia (Björgvinsson, Ehn, and Hillgren 2010; Beck 2002). Yet there are still challenges in moving participatory design into community settings, where its fundamental value of consensus-driven design comes in contact with the contentious realities of dissensus in communities and settings not governed by the obligations and narratives of participatory design's workplace origins. Issues, as a stage-setting device, provide a way to open the design space without supposing particular solutions, which is a slight shift from early participatory design in which the democratizing of design was in the context of developing specific industrial technologies attuned to the preservation of particular labor practices. More broadly, though, this orientation toward issues fits under the umbrella of agonism in design whereby consensus on the precise definition of the issue is not the end goal per se, and may in fact be antithetical to the conception of democratic participation. The alternative is to construct the design process so that it is aligned to afford a plurality of views that will not, cannot, be readily reduced or coerced into agreement, but to create a productive discourse around those divergent views (Mouffe 2005). Björgvinsson et al. (2010, 48) dub these settings "agonistic public spaces," where a "polyphony of voices and mutually vigorous but tolerant disputes among groups united by passionate engagement" are to be heard. More recently, the role of agonism in design has been shaped by DiSalvo's (2012) notion of adversarial design, in which he outlines three tactics—revealing hegemony, reconfiguring the remainder, and articulating agonistic collectives—that describe a design practice for productively engaging in settings around contentious issues. These tactics draw out the relationship between issues and vested groups in different ways, focusing on design as a practice for articulating and participating in a political discourse in which issues, not solutions, are the stock in trade.

Attachments as an Organizing Force

Where issues do the work of identifying the actors, artifacts, and institutions affected by a particular set of conditions, *attachments* do the work of organizing those same constituencies around their shared relations. As Latour and Stark (1999, 24) put it, attachments are "the formidable proliferation of objects, properties, beings, fears, [and] techniques that make us do things unto others." And just as issues configure social design engagements so as to break away from the dualism of management and labor, attachments likewise move us away from the dualism of individual and structure, making room for the messy multiples of relations and motivations that prompt communities to action.

The relationships captured by attachments cross the social and the technical, working at the level of distributed networks of people and artifacts, communities and institutions, and instruments of authority and power. Importantly, the kinds of relationships that the term *attachments* is meant to evoke are not limited to traditional social or political fault lines—e.g., right-left, religious-secular—and instead highlight allegiances that do not, on their face, appear as natural collaborations. Examples of these kinds of attachments are common in issue-based (as opposed to ideologically based) political action. As noted earlier, opposition to genetically modified organisms crossed traditional political lines, with diverse groups of stakeholders, with concerns ranging from globalization and environmental sustainability to civil liberties, all coming together to oppose the development of these altered sources of food (Dryzek 2009; Latour 2007; Latour and Weibel 2005). Similarly, the early development of electric cars eventually brought together diverse stakeholders in advocates of public transportation and car manufacturers (Callon 2004; Callon 1980); open-source software, and more recently open hardware platforms such as Ardiuno, have brought a wide range of individuals together with motivations that are by turns capitalist and anticapitalist, libertarian and communal (Mockus, Fielding, and Herbsleb 2000; Lindtner 2011). In each of these different venues, the visibility of issues becomes a catalyst for enrolling very diverse, and at times ideologically agonistic, stakeholders; their attachments to the issues create an opportunity for working toward a common end, for creating a new public that did not previously exist—one that is defined not just by the makeup of its human constituencies, but by the participation of specific technical and material artifacts that are its raison d'être and the tools by which it does some portion of its work.

Marres (2007) argues that attachments are a series of ontological associations that give shape to an actor's participation and involvement in a set of issues. Central to this involvement, and the motivating association, is the interplay between "dependency on" and "commitment to" the actors, artifacts, and institutions orbiting a particular set of issues. Marres's formulation of attachments as a dependency/commitment dyad builds on earlier work by Gomart and Hennion (1999), who sought to characterize the relationship between different kinds of "addicts"—music lovers and drug users—and their substances of choice. By looking at different forms of addiction, forms at opposite ends of social and moral acceptability, Gomart and Hennion keyed into the tension between action and inaction, as music and drug lovers conditioned themselves to their intoxicants of choice. More generally, this tension between action and inaction gives us a means of examining how sets of social conditions can enable different attachments as actors, artifacts, and institutions come under the effect of a particular set of issues, and then how they begin to enlist themselves to take action to overcome, mitigate, or amplify the effects of the issue of concern. The ontological associations, then, unfold to reveal how a particular set of issues affects the common if incomplete models of experiencing and interpreting the world. These models may be experienced as specific political motivations—environmental sustainability, notions of self-efficacy in contemporary libertarianism, or a sense of common social responsibility for all members of society. They may come through specific arrangements of artifacts—the way cars shape our encounter with space and time, the way computing, and social media in particular, make durable temporary social conditions, or the way those same technologies constrain interaction even as they create novel and more pervasive connections. Finally, the institutions on which we have come to depend reveal these ontological connections—the structures of social service, the organization of political and civic engagement, and the production and consumption of all manner of commercial relations.

The key point here is that each of the actors, artifacts, and institutions (and any other category one might choose to include) participate actively with each other through their associated attachments (Callon 2004). It is not simply that actors produce effects, or that artifacts produce effects, or that institutions (née structures) produce effects; it is that the ways in which effects are bound through their attachments change, evolve, or become otherwise designed. As issues become articulated, the network of real or potential attachments changes. This is, to some greater degree, the work of design, in that we engage in design broadly to create some kind of change in the world through the production of new things. As a method, and in the context of building out community-based technologies, social design provides direct means of addressing attachments as sociotechnical relations that are not fully known but which may be identified through the design work.

By turning to participatory practices to articulate and explore attachments, we develop an alternative to relying on the notion of "frames" to give shape to the design space and indicate which actors might productively participate in the design of new sociotechnical interventions:

The notion of frames stands out as an empirically useful concept to describe how public concern about issues is regulated by substantive means; that is, through issue definitions. … Frames are usually characterized as relatively stable entities—established ideas, values, symbols or institutional devices—that are relied upon to set limits for unstable things. However, a distinctive feature of associations that are highlighted in public issue definitions is that they can no longer be taken for granted. (Marres 2007, 774)

In relation to social design, increasing or supporting participation, on its own, is an act of framing in which the inclusion of different voices changes the frame. Framing, however, is divorced from the issues themselves (it is a way of viewing existing issues), and frames are often taken for granted (as a priori points of view) as "contexts for routine behavior" (Dorst 2015, 65). Dorst does point out that frames can establish a network of individuals and organizations who take on problem setting and problem solving; however, even as trained designers might resist the routine behaviors a frame provides, it is important to point out that the organizations and institutions imbricated in a frame represent particular views, often those connected to professional design practice and product development. Moreover, Dorst's interest in frame setting is in the context of product development: there is a fixed end point at which designers deliver a solution, ideally a complete solution, to a client. The aim of designing publics focuses not on product development but on cultivating new capacities to act beyond an acute problem (I develop this point more below and in the chapters to come). Frames alone do not expose the tensions present in the attachments of a public, because the dependencies and commitments that comprise those attachments are marshaled and modified by the very constitution of the public—they are always inside the public. That is to say, a public exists apart from existing institutions and so cannot be adequately represented by the perspective of those same institutions (Dewey 1954). Attachments, however, do provide a means of understanding the conflicts inherent in the constitution of publics by recognizing the interplay and emergence of dependencies and commitments that form as a public forms: "By approaching issues as particular entanglements of actors' attachments, it becomes possible to credit these entanglements as sources and resources for enacting of public involvement in controversy" (Marres 2007, 775).

Whereas frames can be argued to reinforce these entrenched authority structures, the notion of attachments and publics enables us to move beyond a response to known relations in existing authoritative structures, toward a means of understanding and expressing dynamic authority structures. Attachments, then, foreground the dynamic relationships formed around issues and provide social design programs an alternative narrative for confronting authority and marginalization found in participatory design (Beck 2002; Shapiro 2005). Instead of focusing on the structures of power and the attendant dualisms, attachments move the focus to the broader network of relations and how that totality responds to and produces the issues at hand. One of the early design workshops I held with social service providers accomplished some of the work of articulating their attachments to their social and material practices. This included mapping how their resources—programs, counseling, financial aid—were configured to address particular needs; how each of the service providers was interconnected through referrals, complementary services, and the practices of documenting and sharing information; and how each provider connected these resources, practices, and information schemes to support progress toward sets of common goals. As each organization articulated its own practices with respect to its resources, goals, and information management, it was placed in a larger context that connected the dots between organizations and highlighted how providers were deeply connected through a series of attachments that ranged from instrumental (in their access to hard resources) to affective (in their missions to support personal connection among those they served).

In this way, attachments are both a resource and a goal for social design. As a resource, attachments provide a point of departure, so that as an issue is articulated and explored, there is a network of resources through which to begin understanding, interpreting, and analyzing the condition at hand. As the design process develops, as the public comes into greater relief by exploring potential solutions, the creation of new attachments becomes a goal that alters the public to include a set of resources it did not previously have. In some cases, these networks are obvious and plainly grasped. In others, the opportunities for creating new attachments stem from changes in the way present resources are made available or perceived. In this way, attachments should not be reduced to a set of rationalized properties of actors, artifacts, and institutions, but instead need to be considered for the way they operate and are made available as affective resources—those resources that augment or diminish capacities to act (Deleuze and Guattari 1987). It is the commitment to and dependency on the emotions, beliefs, and desires of publics that make attachments a powerful resource for shaping design and for providing an environment within which designed interventions might be sustained over time.

Infrastructuring: Designing Sociotechnical Futures

Publics, the issues around which they form, and the attachments to diverse resources upon which they draw provide a conceptual scaffolding for understanding forms of community action that revolve around marshaling diverse resources to confront social issues. As a public identifies and marshals the social and technical resources to contend with social issues, these resources become a form of infrastructure for the public: a durable and ready-to-hand support that enables constituents of a public to act. The work of creating it is a process described succinctly as *infrastructuring*, in which the infrastructure arises out of the relations and the resources entangled in the present issues and attachments.

The turn toward infrastructuring (a term borrowed from Star and Ruhleder 1996 and Star and Bowker 2002) grows out of a recent move in design research aimed at examining how we might return to the principles of Scandinavian participatory design when working in community contexts (Ehn 2008; Binder et al. 2011; Björgvinsson, Ehn, and Hillgren 2010). Rather than approaching participatory design as a method of product design focused on responding to present conditions, Ehn and others would configure it as a kind of metadesign:

Hence, there will be a shift in focus from design-games aiming at useful products and services, to design-games to create good environments for design-games at use time. Typically this will at project time lead to an occupation with identifying, designing and supporting social, technical and spatial infrastructures that are configurable and potentially supportive of future design-games in everyday use. (Ehn 2008, 96)

The idea of infrastructuring through design turns on the distinction between a primary concern with design-for-use—centered on end products—and a focus on design-for-future-use—intended to create fertile ground to sustain a community beyond the development and deployment of any one specific artifact. This entails a shift from treating designed systems as fixed end points to treating them as ongoing infrastructure: sociotechnical processes that relate different contexts (Star and Ruhleder 1996). Attending to design as a process that continues through adoption and use—a process that grows out of capacities developed by engaging in co-design—shares affinities with a larger set of arguments in the design literature around the similar process and role of metadesign in building capacities for action (e.g., Fischer and Giaccardi 2006; Fischer and Scharff 2000). With respect to publics, however, infrastructuring is the work of integrating sociotechnical resources—via existing and newly articulated attachments—that enable adoption and appropriation beyond the initial scope of the design space (Björgvinsson, Ehn, and Hillgren 2010, 43):

Infrastructuring entangles and intertwines potentially controversial "a priori infra-structure activities" (like selection, design, development, deployment, and enact-ment), with "everyday design activities in actual use" (like mediation, interpretation and articulation), as well as "design in use" (like adaptation, appropriation, tailoring, re-design and maintenance).

The distinction between design *for* products and design *as* infrastructur-ing invites us to broaden our view of what counts as invention. The move here is from a technocratic view of invention toward one that includes social invention—which is to say, invention that arises from the constitution of a public, grounded in issues and born of attachments. One could express it as the difference between federating individuals as a multi-stakeholder response to known issues (in the case of product design) and constructing constituencies as previously unknown issues are discovered (in the case of infrastructuring). It is not the product that matters but the development of certain practices and ways of relating to issue formation and the articula-tion of attachments, which can lead to the productively democratizing ele-ments of participatory design such that participants are empowered beyond the scope of the design process. What Ehn and others have argued for rests on an understanding that technologies, once deployed, take on a life of their own—both as cause and effect, they participate fully in the social con-text (Callon 2004)—and if we are to go back to the early motivations of participatory design as a politically grounded intervention to empower the disempowered (Beck 2002), then the most effective way we might do so is not through the production of new technological artifacts per se, but by scaffolding ongoing collaborative design practice: focusing on "design for future-use" instead of for "present-use" (Ehn 2008), embracing the socio-technical as a dynamic resource rather than as a stable frame within which technology is deployed.

To illustrate this point, we can turn to the literature on public partici-pation and its many examples of computing systems designed to engage and support different forms of participation. Several projects have used computationally mediated participation to bring diverse stakeholders into discourse over shared policy and land development decisions (e.g., Borning et al. 2005; Friedman and Kahn 2008; Hirsch 2010); community-wide tech-nology deployments have been developed to support information-seeking practices and sustain civic participation in various kinds of communities (e.g., Hampton and Wellman 2003; Carroll and Rosson 2003; Carroll and

Rosson 2007; Jackson et al. 2004; Pinkett and O'Bryant 2003); and new forms of activism are being explored through developing perspectives on the urban environment and ways in which new forms of technology can support participation, public discourse, and modes of reenvisioning the city (e.g., Bayea, Geith, and McKeown 2009; Carroll and Ganoe 2009; De Cindio, Di Loreto, and Peraboni 2009; Klaebe et al. 2009; Veith, Schubert, and Wolf 2009). These projects are examples of public encounters that raise important questions of how to build large-scale participatory systems. Each of these lines of research makes claims about how its practitioners were or were not successful in developing meaningful participation and sustained engagement within the communities that they developed or deployed their interventions. A useful way to disassemble these successes and failures is to consider to what degree the work of infrastructuring occurred within the contexts of development and deployment. The answer to this question depends not simply on whether or not the final artifact was adopted and used, whether the intervention had some point-in-time effect, but on the degree to which it altered the sociotechnical landscape. The impact of technology deployments ranged from little or none due to a dearth of social reciprocity (Jackson et al. 2004), to successful community uptake when supported by active local facilitators (Pinkett and O'Bryant 2003), to such strong community engagement that the residents took action against the construction company that had built their homes (Hampton 2010). Where these projects sustained publics, we find infrastructuring to be both social and technical: key community actors facilitated and supported the new technological capacities, and the attachments between community members grew out of the collective participation of people and artifact as they worked jointly to address particular shared issues.

What infrastructuring does, then, is move from a focus on creating a particular artifact (and the attendant fixity of context and artifact) to design as constituting a public in which issues and attachments are conjoined into sociotechnical networks for addressing present and future conditions. Moreover, infrastructuring plays a particular role in the constitution of a public by making durable the relations and resources that enroll affected actors and artifacts and institutions: articulating issues and attachments are both necessary preconditions for constituting a public. The final active ingredient is the process by which a public internalizes those issues and attachments and operationalizes them as sociotechnical capabilities for contending with present and future conditions.

Social design, building on a legacy of participatory design, possesses the epistemological machinery to contend with the political and agonistic challenges of constituting publics, of building out new capacities by enabling designers and community members to eschew entrenched stakeholders and known issues and to explore a breadth of possible outcomes through alternative articulations of shared social conditions and novel allegiances. Indeed, a move toward approaching social design as a matter of constituting publics, rather than products, is consistent with a reformist or activist agenda of broadening the impact of participatory design beyond the production of products toward the empowerment of the disenfranchised (Shapiro 2005). The act of infrastructuring is the core to supporting such an agenda as it moves past participation as a framing for design toward participation as an ongoing act of articulating and responding to dynamic attachments. The public, however it might be constituted, is a sociotechnical response to these dynamics.

Social Design and the Case for Publics

Latour and Weibel (2005) make an early case for publics in design in *Making Things Public*, in which they present a collection of projects that provide a platform for the collective expression of, and response to, contemporary sociopolitical conditions. The examples in that collection highlight the connections between issues, attachments, and infrastructuring that constitute publics by providing evidence of the multiple ways actors, artifacts, and institutions become involved in sociopolitical conditions. These diverse constituents act on each other, and are neither the sole causes nor effects of an object-oriented democracy in which objects are acknowledged as playing a vital role in the production of publics as people (Latour and Weibel 2005; Callon 2004). The broadening of the field of view to include objects in the exploration of political actions, and in the notion of what constitutes a public, is important because it provides a perspective from which to engage with the attachments to issues, and exposes the socio technical infrastructures that shape how a diverse set of actors might take action.

All of this—the issues, the attachments, the work of infrastructuring, and the constitution of publics—is crucial to the development of social design as an explicit agenda within design, whether design concerned with digital technologies and interactions, or projects undertaking public or commercial service design, or even traditional forms of product design where artifacts are part of how we organize and understand participation in the politics of environmentalism, sustainability, and consumerism. Just as we turn to "social justice" to encompass a great many ways of acting in the world toward the betterment of those in need, without voice, or otherwise without the resources to act effectively on their own, so social design is meant to evoke a narrower set of actions we take through the creation of systems and artifacts that are meant to be placed into the world to empower, support, and act as resources for individuals and groups for whom such things might not normally be available. Social design is about addressing the needs, desires, hopes, and aspirations of people by facilitating the creation of imagined futures and providing the tools for acting on those imagined futures. Such a design agenda means working directly with social conditions and taking on political, moral, and affective modes of interpreting and acting in the world. It means techniques are needed that confront entrenched views in order to create alternative futures that build persistent and transformative social and technical capacities. A public puts into relief those actors, artifacts, and institutions affected by a common set of issues. Social design—as developed here and drawing on participatory design's legacy of engaging in politically charged settings—provides a set of methodological tools easily purposed for engaging with diverse actors with potentially divergent goals in mind.

The chapters that follow detail a case study for designing publics as a framework for social design by examining an extended technology design engagement with providers of homeless services. I have previously reported on parts of this work in other venues, to which I will refer; here I limit my retelling of the ethnographic account because my focus is on the larger arch of how a public emerged and took shape. The analysis provided here is synthetic rather than piecemeal; it builds on my earlier ethnographic accounts but has its focus fixed on a broader landscape rather than the minute details of particular features within that landscape.

By way of synopsis, beginning in mid-2007 I surveyed the local landscape of homeless care provision and the multiple experiences of homelessness that individuals and families confronted between a rapidly shifting economic landscape and the lingering aftermath of Hurricane Katrina and the regional displacement it caused. Over a year of qualitative and ethnographic fieldwork enabled me to gain insight into both the conditions of homelessness (Le Dantec and Edwards 2008a) and the conditions of social service provision across a portion of the larger nonprofit ecosystem in Atlanta (Le Dantec and Edwards 2008b; Le Dantec and Edwards 2010). From that early fieldwork I developed a longitudinal design engagement, first running participatory design workshops with care providers—individually and collectively—and then focusing my work within the shelter described briefly in the previous chapter.

As I transitioned to participatory design activities at the shelter in early 2009, I would work with the staff and residents of the shelter on a weekly basis to create, refine, and test different ideas for how information might be shared, how the work of connecting to services might be organized, and how the implicit social support network of the shelter might be made visible and durable given the rapid turnover of residents (and to some degree of staff). Over the next twenty-four months, I worked with six different staff members and 38 families through weekly design activities. These activities ranged from small focused work to settle a particular feature, to larger problem scoping, and they led to an environment where we were solving a range of problems together regularly. From the daily challenges of coordinating chores, to the longer-term interventions and follow-up needed to help create stability and self-sufficiency in families struggling through difficult economic and personal crises, the design work enabled the mothers and the shelter staff to imagine new ways of working in the world and new ways of relating to each other in that work. The participatory design engagement created moments of empowerment as the mothers became the architects of solutions to their own issues. It also enabled the staff to examine the roles and relations within the shelter such that the mission of the shelter became more overtly collective as the technologies we designed together made a network of information and personal connections more visible (Le Dantec et al. 2010; Le Dantec et al. 2011; Le Dantec 2012).

It was through this design engagement that I established the empirical basis on which to explicate the relationship between publics, issues, attachments, and infrastructuring.

3 Articulating Issues

In order to build out how publics form in more detail, I first turn to *issues* and how they were articulated and formed through a series of design interventions I ran with care providers in the context of urban homelessness. The importance of issues, together with attachments (which I will treat in more detail in the following chapter), is the way in which they can be used to destabilize assumptions about relevant stakeholders. This makes issues a useful point of departure and resource for social design interventions precisely because it is often the entrenchment of stakeholders and their specific positions of authority or power that need to be overcome, or at least perceived in different ways, in order to move forward toward novel technical or social solutions. By stepping back from aligning with a particular group or established position, and instead beginning by articulating issues, community-engaged design interventions can then create a discursive space in which multiple views can come together.

Marres makes the case very well for issues as a way to break from established frames when she writes that "frames are usually characterized as relatively stable entities—established ideas, values, symbols or institutional devices—that are relied upon to set limits for unstable things" (Marres 2007, 774). There is a parallel here between frames and stakeholders, as they are both meant to provide a stable basis for investigating or otherwise contending with the world. When taken as absolutes, stakeholders act as a kind of matter of fact (Latour 2008; Latour 2004), immutable and unassailable as a potential site for contesting designed futures. This immutability is not useful, as it imposes dualisms analogous to what Latour critiqued when trying to reconcile a social reality placed in reference to a material reality (Latour 2008).

Latour's critique traces the alternative epistemologies of science and art and the historic imposition of an empirical bifurcation in which we have "on the one hand, a harsh world made of indisputable matters of fact and, on the other, a rich mental world of human symbols, imaginations and values" (Latour 2008, 39). For Latour, matters of fact arise out of this historic split that separated the material world, made up of primary qualities of objects and soundly the domain of scientific endeavor, and the symbolic ordering of the world from which art arises. Working with philosophy—which has traditionally been the discipline tasked with resolving the material and symbolic worlds—Latour moves to reframe this imposed epistemological division between facts and experiences by tracing through the radical empiricism of William James and the speculative philosophy of Alfred North Whitehead, arriving at the concept of matter of concern:

A matter of concern is what happens to a matter of fact when you add to it its whole scenography, much like you would do by shifting your attention from the stage to the whole machinery of a theatre. … Instead of simply being there, matters of fact begin to look different, to render a different sound, they start to move in all directions, they overflow their boundaries, they include a complete set of new actors, they reveal the fragile envelopes in which they are housed. (Latour 2008, 39)

For our purposes here, an analogous binary occurs not between material and mental states, but between present and future states. Design becomes the means of reconciling the two (Simon 1996), bridging the riverbanks and leading us to a preferred condition. Yet the problem for us, as it was for Latour, is that neither side of this riverbank is fixed, nor does design act as a stable passage between the two. Instead, design alters both through its very act. So while the idea of stakeholders is seductive because of their stability, by reducing them to a collection of fixed positions in order to treat them as matters of fact, their agency in the larger situation is largely bracketed. However, it is precisely the ways in which stakeholders act in the world that are the most relevant for constructive community engagement. It is, as Latour argues, the contextualization of matters of fact, creating matters of concern, that is most important here: stepping away from the given, the stable, and instead looking at the ways different actors and artifacts and institutions act in, and relate to, the social and technical situations in which they find themselves.

By focusing on issues as the point of entry for a social design intervention, we create alternative ways of challenging the stability of stakeholders as natural allies, and instead focus on the matters of concern that enroll actors and artifacts and institutions as participants. Issues provide a perch from which to develop a rich understanding of the context and of the current practices employed within that context. As I developed in the opening chapter, this is an important first move for community-based projects because the dynamics present are multiple and complex in ways that are different from design in the commercial workplace, where interaction design and human-computer interactions have established methods and a legacy for operating within the confines of commercial production.

In the context of community-based design, it is not always clear where antagonism originates—and a single point in which to locate the source of a shared issue is unlikely. This very lack of easily traceable connections and of the downstream visibility of the causes of social conditions is precisely the problem that Dewey's public was trying to contend with: it was that "the machine age has so enormously expanded, multiplied, intensified, and complicated the scope of the indirect consequences" that taking action on an issue was made difficult, if not impossible (Dewey 1954, 126). So to focus on issues in constituting a public is to begin to trace the shared social conditions, rendering visible the direct and indirect consequences of those conditions across diverse actors and artifacts and institutions, allowing the issues to inform and guide where and how these actors and artifacts and institutions might be involved in, and evolved through, a design agenda meant to confront the issues.

Issues: Outside And Inside

When issues are approached as the primary framing activity, two important connections come to the fore. First, privileging issues aligns with a pluralist view of "wicked problems" in design by providing a common context for approaching and enlisting diverse actors (Buchanan 2001; Dorst and Cross 2001). Second, issue formation and the work done to articulate connections and consequences of those issues places a focus on the multiple practices that may be contending with the issues in a given context (Law 2004). To the first point, the wickedness of the problem is not simply about giving shape to the design space, but that the inclusion (and more often exclusion) of different voices, perspectives, and values comprehensively changes the way such figurative spaces are constructed and navigated. In Dorst's words (2006), this leads us to understand that the "ill-structuredness of a problem may not be an *a priori* property of the problem itself, but is linked to the capabilities of the problem solver." As issues circulate as matter of concern, flowing between contexts, they interact with different actors, artifacts, and institutions whose backgrounds, motivations, value structures, and ontologies vie with each other, producing a set of wicked problems and a socially constructed design space.

The second point builds on the first: as issues circulate through different contexts, morphing from Latourian matters of fact to matters of concern, they also circulate through a complex set of local cultural and professional and disciplinary practices. This confluence of multiple perspectives that develop through a shared set of issues is akin to what Law termed *multi-sitedness* (2004). For design to engage in these settings in which multiple practices are entangled in issues, it needs to confront and navigate a complex set of locations where these practices unfold. Furthermore, it is not just that design engages with the mess—with complex issues—but that it also actively participates in enacting those issues through methods of intervention, generation, and resolution.

In the world of homeless care and social service provision, the institutional contexts in which issues circulate include a regulatory structure of how and to whom services are provided, the largely nonprofit apparatus of providing those services, and the individuals and families who receive or make use of those services. At each node in this network—one that spans national, state, and regional policies, along with a diverse cadre of service organizations, and an even more diverse population of people seeking help—the issues at stake manifest themselves and are interpreted differently.

For the regulatory structures, the issues are largely sited within resource allocation and accountability. There is an express need for accurate information, in part to ensure compliance with existing regulations but also to enable shifts in policy or priority to address social needs as they arise. The issues revolve around the ways in which data are captured and categorized, and the deployment of technologies meant to support the implementation of public policy has largely focused on those instrumental needs (Bardram 1998). When viewed from the regulatory structure, the issues manifest themselves as outside the particular problems of those seeking social services, but they need to be addressed in order to assess the impact or efficacy of public policy.

An example of the issue of regulatory compliance and its role in motivating changes in social service practice comes from the rollout of the mandated-use Homeless Management Information Systems (HMIS). In 2003, Congress mandated, via the department of Housing and Urban Development (HUD), the collection of service data in electronic form (Sarpard 2003). Prior to that mandate, there was no systematic national data collection for social service programs. The issue of data collection for HUD is arguably outside the issue of homelessness or of providing service to the urban homeless. It is, in practice, simply an accounting of services, one that is about enforcing compliance with existing regulation. It is also aspirational in enabling the application of analytics to the outcomes of different kinds of services provided and in enabling productive data sharing between service providers to improve service provision. The challenge, of course, was that these issues, while appealing to regulators, were not aligned with the issues experienced by care providers nor by the individuals and families who depended on those care providers.

For local nonprofits providing services, the issues they primarily faced centered on making their limited resources go as far as they could to serve the communities in which they were situated. Accountability was a component of resource management; however, accountability was an obligation to be met in order to continue to have access to resources and not necessarily a tool for service improvements. As an example, the categories of service present in the HMIS system often did not match individual circumstances and the contexts in which services were being provided, so caseworkers had to account for their work in two ways, once for the regulatory system and once for their own specific organization. This kind of "off-books" accounting of social services, in which individual providers made case-by-case decisions and chose how to best document those decisions, was how they managed-up to regulating bodies while maintaining autonomy to respond to local issues (Lipsky 1980; Le Dantec and Edwards 2010).

The most intimate level of experience of social services comprises the encounters individuals and families have while seeking assistance. The issues that arise here revolve around the visceral day-to-day realities of being homeless. These realities require the development of tactics to identify and make use of what few resources are available and to learn how to navigate a support system that is very focused on trying to alleviate dire individual circumstances through an often didactic and normative approach to engaging with social problems. The issues of regulation are even further from what the homeless contend with, even if issues of regulation have direct, though often invisible, daily impact.

The result is layers of issues and consequences that are enacted across a multitude of sites. The challenge for design and the constitution of a public is to mediate these multiple sites, to constitute a collective response that engages rather than obscures the mess and multi-sitedness. Reconciliation is not necessary, but by focusing on the practices present at each of these nodes in a larger network, it is possible to tie together solutions that work with the plurality introduced by these different scales of issues.

While the issues manifest themselves differently, there are important ways they overlap. To begin to meet the challenge of issue formation at each of these sites, I developed a series of participatory workshops. Starting with issues of service provision, I worked with care providers, HMIS developers, and local regulators to articulate the unique and common issues that circulated among social service organizations. What resulted was a set of shared issues that existed "outside" or at the institutional scale of homelessness: needs served, resources available, service goals, information flows, and accountabilities to external institutions (public and private). Following the articulation of issues "outside," I then turned my attention to a single provider in order to articulate issues that circulated "inside"—or at the local scale—where homeless families and caseworkers interacted to overcome immediate housing crises. Through the fieldwork and initial design interventions, I am able to trace how common issues emerged and were enacted at different sites and the ways they were made visible and material across the spectrum of institutions and people enlisted in the larger context of urban homelessness.

Institutional Issues through Design

The institutional scale, as I am defining it here, follows issues from the level of regulation down through the local service provider. As pointed out in the beginning of this chapter, in my study these issues were based on contending with homelessness in the aggregate, in which programs and services were viewed as part of a larger system of social support and where resources needed to be managed across large and diverse populations. At the level of regulation and policy compliance, issues were largely based on concerns about managing resources across diverse sets of providers. The local service providers were included in the institutional scale because some important measure of the work they did on a daily basis was in service to the needs of regulators. The local service providers also inhabit the local scale, and, as might be expected, much of the way they provided direct service came down to the way they translated and interpreted between these two sites (Lipsky 1980; Le Dantec and Edwards 2010).

In order to articulate the issues at the institutional scale, I brought individuals and administrators from different care providers together for a series of design investigations. The investigations began with interviews and fieldwork at each of the eight different agencies with which I was working. The service providers offered a range of services including legal aid, housing, health care, direct financial aid, employment readiness, and addiction and drug counseling. The initial fieldwork structured my understanding of how social services were provided in the city (Le Dantec and Edwards 2010; Le Dantec and Edwards 2008b), and exposed the kinds of issues facing each agency within its own context of operation. Following the initial fieldwork, I hosted a workshop with sixteen administrators and social workers representing each of my partner agencies. I brought them together specifically to understand the interconnections within the larger ecosystem of homeless social service provision. The aim of the workshop—and of the follow-on fieldwork—was to conceptually map three key areas that informed their work practices: first, the resources available through each of the care providers, including the appropriate audience for receiving those resources (the public at large, other caseworkers, or homeless clients); second, the goals that each resource contributed toward, including detail on whether the goal was a response to the needs of those being served, or existed in the context of structuring services in the community; and third, the flow of information at the care provider, including information about clients, information shared between providers, and information necessary for external regulators. These three elements—resources, goals, and information flows—were the kinds of issues that surfaced at each individual agency, but that also circulated in useful ways across individual providers and across scales of involvement in the larger social service ecosystem. By attending specifically to the way resources, goals, and information flows

Figure 3.1
A sampling of the conceptual maps that were created during early design workshops.

were enacted and responded to across the eight sites, I was able to create a series of conceptual maps that traced how issues became matters of concern as they moved out of native contexts of interpretation and action (see figure 3.1).

The maps, created through a series of diagramming and documenting exercises, helped make the *whats* and *hows* and *whos* of service provision visible across the organizations by linking each provider's goals for the programs they ran, the available human and material resources for achieving those goals, and how different services and service providers were connected through different practices of sharing information. By building conceptual maps around the work practices at each provider, the individual issues of managing resources to provide services, collecting and using data for counseling and client interventions, and meeting obligations for institutional reporting all became entangled in an ecology of resources, people, and data that moved between different care providers and was represented to each in different ways. What were matters of fact at one provider became matters of concern as the group recognized the ambiguities, tensions, and misalignments of service provision that occur as people and resources were moved from local contexts into the larger network. Within this frame, the two primary issues that were articulated were those of *managing data* and *managing relationships*. Data management was expressed through the information flows and the ways that goals and resources were structured, delivered, and communicated outwardly to different constituents. The issue incorporated elements of work practice, as well as data representation across different kinds of accountabilities and obligations. Relationship management included both the relationships with the individuals being served by each of the organizations, as well as relationships between organizations. The challenge that was expressed through the workshop and made visible by the conceptual map was that keeping track of the *whos* was contorted around a mismatched set of *hows* as information flows were divided and duplicated across multiple channels. These multiple channels of information management became the foundation for the issue of managing data, because data about *whats* and *whos* were collected and kept in multiple kinds of systems.

Data Management

At root, the issue of data management articulated by the care providers was tied to particular problems around the *hows*, which is to say the different data management practices at each agency. These data management practices were developed as local responses to the mandated use of a common HMIS system, a system that had its own history within the context of supporting the provision of social services in the city. This history, in turn, meant that the representations of services in the HMIS were neither adequate nor appropriate for many of the kinds of activities undertaken across the very diverse spectrum of social service providers that operated in the state. Elsewhere I have documented some of the details of how this system arose as the technology of choice for the state (Le Dantec and Edwards 2010). What I would like to point out here is the way the issue of data management became a matter of concern as it moved between the specific manifestations at local providers and the more general concerns of regulators and state-level oversight. In the workshop, the service providers and agency directors were able to trace their own local data management practices through the larger common network. While each of the participants came into the day with some understanding of how they fit within that network, the specificity of the activities in cataloging the resources, goals, and information flows enabled very specific reflection and realization about how their data management issues were bound to parallel practices at other providers. For example, three of the organizations worked with a similar population of chronically homeless men; through the workshop activities each realized they were using the HMIS differently, which meant local data management choices at each of the providers were creating redundant work simply because there was no agreed-upon common practice for capturing and sharing client information. By tracing this issue through different local settings, it became possible to map out the initial contours of a public through the way different actors and artifacts and institutions were entangled in the consequences of these interlinked actions. Where the individual provider might have developed very specific strategies for contending with the issue, systemically the certainty in how to overcome the issue dissipated as the issue overflowed its boundaries (Latour 2008).

One of the main concrete issues that arose out of the design workshop involved the procedural need to generate reports to funders—both state and federal as well as private foundations. This one issue, the generation of reports, captured the limitations of the HMIS. The kinds of data that were relevant for managing bed counts for homeless single men were different from those needed for families; moreover, the kinds of data captured were different for shelter bed management than they were for the many other kinds of services—health care, food assistance, employment readiness, financial aid, disability assistance. These differences meant that many of the local agencies had to develop separate modes of data management in order to support their local practices of social service provision, while duplicating some work and some data for the purpose of reporting through the HMIS for regulatory compliance. The driving force behind using the HMIS was the need to record client data to generate reports to funding bodies—typically government-based funders at a combination of municipal, regional, and state levels. However, the transformation of the HMIS from a case-management-focused system for shelter bed sharing to one used for policy compliance meant that the key reports necessary to demonstrate compliance were often difficult to generate, because the initial assumptions about which data were relevant and how those data would be entered did not match the kinds of transactions recorded by caseworkers at the many different sites where the system was in use.

To illustrate this, to preserve the privacy of homeless individuals, the HMIS was designed so that reports were built from service transactions and not from individuals entered in the system, even though both sets of data were present in the HMIS. Compounding the problem for service providers was that many kinds of services that some of them provided were not present (or not usably presented) in the HMIS. The result was that caseworkers, unable to select an appropriate service transaction, would simply enter free-form notes about the services a person or a family was receiving, attaching that information to the record for the individual. Then, when it came time to generate reports, agencies found themselves substantially underreporting because many of the individuals they served did not have service transactions associated with them and were therefore omitted from the final report.

The original data model that used the service transaction as the atomic unit for accountability grew from a protective instinct to turn the attention of oversight away from a vulnerable population comprising individuals with mental health issues or long-term struggles with addiction, and who are contending with the consequences of multigenerational poverty (Beegle 2003). Rather than individuals, the motivation was to track the services rendered and the organizations that provide those services. (The usual phrase was "Don't track the person, track the service.") Unfortunately, this priority in practice frustrated many of the goals that regional HMIS rollouts had been intended to meet. Generating accurate reports was only part of the problem; the more significant mismatch was not about compliance but about building a more robust and sophisticated care management scheme across a region, one that would enable care providers to know and understand the kinds of services an individual was receiving so they might make the most impact for that person.

There are two ways to view this issue. One focuses on efficiency in providing resources, in which the goal is to remove duplication of services and flag potential abuse. The second is to view such tracking as a way for informed providers to make better judgments about what kinds of services will have the most impact based on what the person has received recently. The challenge, of course, is that these perspectives on the issue coexist. At the institutional scale the motivation is very much put into the context of efficiencies, compliance, and mitigation of abuse; however, at the local and situated scale where an agency or coalition of agencies want tools to better serve their communities, they want better support for coordination and informed decision making. The issues overspill their boundaries in both directions and are mixed with varying degrees of wariness of the intent from the distant perspective.

Relationship Management

The second issue to discuss is that of relationship management. This issue was closely tied to data management, but instead of growing out of the need to track and instrument services provided, it was based on the direct interactions of caseworkers providing those services. The issue was equal parts *who* and *how* and bore the weight of creating movement through the conceptual map created at the workshop. Where the goals, resources, and information flows marked out kinds of locations, way points, and paths of arriving—and highlighted the challenges of getting from one location to the next in light of which data were managed and how—the creation and management of relationships was the engine that moved people through the service space.

For the caseworkers, the practices they were developing in conjunction with the HMIS revolved around data capture and tracking required for delivering evidence-based outcomes. Since many of the features needed to reach these goals were not available directly in the HMIS—or were present in ways that made their integration into daily work practices onerous and unlikely—caseworkers and staff had to maintain parallel systems and processes that would enable them to effectively manage their resources in service of the goals they were developing with their clients (Le Dantec and Edwards 2008b; Le Dantec and Edwards 2010). As I noted earlier in this chapter, when the HMIS was developed, its workflow and data model were built around capturing service information with the goal that it could be usefully shared across service providers. The challenge for the caseworkers using the system was that the data management practices built in to protect the privacy of individuals receiving service effectively rendered the HMIS useless as a tool for coordinating services across different service providers. The consequence was that, in addition to developing parallel data management practices to overcome the impedance mismatch between the HMIS and work practice, caseworkers had to rely on social connections as the primary vehicle for coordinating care across the network of providers.[1]

By way of example, because information about clients was carefully obfuscated and partitioned, caseworkers had to go through a multistep and manual process to figure out in any actionable detail the service history of a particular client. The availability of these details was contingent on explicit permission having been granted to view shared information, which meant the caseworker might be able to view only partial records of service transactions or might not see usable information at all. Regardless, these barriers meant that more often than not, when a caseworker needed to know more information about a particular individual, they would have to pick up the phone. Caseworkers then spent considerable time on the phone, not just managing their own clients but providing context about former clients to other providers. While this system of direct contact worked, it was counter to the aspirations of the HMIS, which was meant to address these very issues of information sharing and coordination.

In order to be effective, a caseworker needed to maintain strong professional relationships within the network of providers and create good working relationships with the clients being served: it was imperative to build trust and develop a working relationship in which both the caseworker and the individual being helped could be frank and constructive in assessing how certain programs were, or were not, working. There should be no surprise that social services depend greatly on establishing strong working relationships between a homeless individual and their provider (Hersberger 2003). What this serves to highlight, however, is that social work is, through and through, *social*: whether with a professional network or with clients, care providers had to build and manage relationships in order to successfully intervene in the lives of those who came to them in need. Without strong personal connections between organizations, the ability to effectively connect homeless individuals with services in the larger network was hampered. Without robust relationships with clients, caseworkers' ability to support their progress through programs was also constricted.

As a result, relationship management was an issue with wide-reaching consequences. The connection between the *hows* and the *whos* and the *whats* became dislodged from the particular social issues being addressed: marshaling resources came down to which caseworkers and providers were connected through professional relationships; establishing and working toward goals—individual and institutional—rested on caseworkers building relationships with clients; and generating momentum through the system, through the network of personally connected providers, required connecting and maintaining relationships across multiple boundaries.

The issues of data management and relationship management were intertwined and grew out of the forced marriage between historic ways of managing client information and the mandated use of the HMIS. The practices articulated through the workshop amounted to a constant back and forth between relationship management in a wide network of providers and the capturing and sharing of related data through different channels depending on the task at hand: internal systems for case notes and day-to-day management, program-specific systems for enrollment and benefits, or the HMIS for regulatory compliance and accountability (Le Dantec and Edwards 2008b; Le Dantec and Edwards 2010).

As these issues became clear through the course of the workshop and the conceptual map that emerged through engaging diverse providers in articulating their shared networks, a more comprehensive view of the issues began to emerge: the distant regulatory conditions that sat behind the HMIS came into focus and were situated more concretely with their daily consequences; the influence of community values common among the providers became visible and their consequences related to daily practice; and different experiences and tactics for contending with these issues were shared across providers with different missions, working under different kinds of regulatory regimes (e.g., health versus financial aid). While the workshop did not specifically seek to propose resolutions to the multiple ways the providers dealt with data and relationship management issues, it did serve to make those issues visible by interlinking resources and goals and information flows to provide a concrete way for the eight providers to see how their local practices affected other organizations in the network. Issues as matter of fact regarding how a single provider might deal with data and relationship management became issues as matter of concern through their circulation in the larger network of providers. Moreover, instead of treating the setting for a design intervention as one of disparate stakeholders, each with unique and entrenched needs, the notion of a public began to emerge, one that was connected through these issues and through the daily practices they had developed to contend with these issues.

Local Issues through Design

I want now to turn to the local issues of how social services were meted out to homeless individuals. In doing so, I will begin to draw connections between regulation and its effects on the daily lived experience of individuals seeking and receiving care. The issues that dominated the local scale centered on how to translate the resources of the providers into interventions and outcomes to meet the goals established with those they served. For the care providers, this meant the daily work, not of data capture and management, but of contingent decisions on how best to intervene in the particular conditions of each individual who walked through their door. For the homeless, it was the work of survival in piecing together needed services and aid, and the work of establishing some seed of stability as they tried to look beyond immediate crises. As with the institutional issues discussed earlier, the point to attend to is the ways in which highly specific issues as matters of fact migrate within the larger context to become issues as matters of concern that enable the formation of a public, a site for collective action.

Earlier fieldwork had started to sketch out some of the issues faced by the urban homeless and the specific ways different forms of technology were implicated in those challenges (Le Dantec and Edwards 2008a). These included staying connected to friends and family who may have become geographically distant, managing identity and the way the stigma of homelessness affected their ability to operate in society, and gaining access to accurate and reliable information for services and opportunities to regain self-sufficiency. Bound up in these issues were specific and individual matters of managing personal health and chronic disease,[2] gaining access to and using transportation, finding and maintaining employment, and exercising basic financial literacy. Different forms of technology governed much of their day-to-day lives,[3] mediating how they stayed in contact with their friends and family, how they sought and found information, and how they made use of public transportation. Mobile phones, Internet-enabled services, and electronic fare cards for bus and rail each had particular affordances and consequences for the urban homeless, variously enabling continuity of social contact through the use of mobile phones, creating an illusion of access and richness with online services and search, and stymying budgeting and planning by hiding currency and services behind electronic sensors only accessible at the point of sale or use.

While many of the technologies the homeless came in contact with were viewed with skepticism and frustration, the one that resonated with the individuals with whom I was working was the mobile phone: as a technology, it was instrumental in mediating many of the day-to-day issues they faced, but more importantly, it was technology that was legible and attainable. During my early fieldwork, the homeless individuals I worked with would talk about using a phone to manage personal information, to entertain themselves, and to take and store photos. At the same time, the homeless individuals I worked with did not discuss these activities in the context of personal computers. The value and utility of the mobile phone was apparent. The personal computer remained just out of reach, financially and experientially, and as a result was not a legible technology. As a tool for communication and information seeking, the personal computer was on the far side of the digital divide, a symbolic artifact rather than a practical tool. It is for these reasons that mobile-phone-based technologies became a point of departure for co-designing new modes of interaction between shelter staff and residents.

In order to connect the local issues the homeless were contending with to the institutional issues of the care providers, I cultivated a participatory design partnership with one of the providers from the workshop. The site chosen was an emergency shelter that provided thirty to ninety days of emergency housing to single women with children. As described in the opening chapter, this shelter became the primary site for co-creating a set of technologies and, in the process, constituting a public. I chose the site based on the centrality of the provider with respect to the other service providers present during the workshop—as it provided emergency housing to single-mother families, much of what the caseworking staff did at the shelter was connect residents to the multitude of services they needed to stabilize their lives. My goal was to make connections to the institutional issues that would resonate with the providers and be a resource for design. During the initial six months of an eighteen-month participatory design engagement, I met with the care providers and residents of the shelter in order to develop a design-led program for identifying and addressing issues there. The design work developed over time through group and individual co-design sessions where the shelter staff, residents, and I designed the technical features and physical installation of a communication platform. The initial meetings focused on understanding how the work practices at the shelter were situated around the resources, goals, and information flows expressed in the design workshop and grew to include the social support network among residents. This enabled me to contextualize the issues of data and relationship management with respect to the specifics of how caseworkers interacted with residents at the shelter. When working with the mothers at the shelter, we focused on developing a shared understanding of their particular experiences in gaining access to services, the individual challenges they were contending with, and where and how they sought information and support. I wanted to understand their perspective on what they were doing for the relatively brief time they were at the shelter, what their relationships were to each other and to care providers both at the shelter and at external organizations.

One of the first things that came out of these early participatory design investigations was that while the conceptual maps developed in the workshop were useful as a way to expose resources in the community, the more fundamental challenge faced by both caseworkers and residents was managing communication. For the caseworkers, the issues they were focused on included their need to manage multiple relationships, coordinate actions around service provision, and deal with resource constraints. Shelter residents were focused on structuring the information they received to help with information overload, establishing and maintaining relationships at the shelter, and developing a network for social support. Managing communication was where these issues met as a shared set of conditions that bound the caseworkers and residents together while they worked through individual programs of counseling, and that connected to the institutional issues governing service provision from afar.

Shelter Staff

From a distance, the needs of the shelter staff aligned with issues described in the previous section: they needed to manage data and relationships as part of their work in providing social services to the residents. On closer inspection, however, they faced a number of specific and particular issues, which they articulated through the design interventions and interviews.

The first issue centered on the resource constraints placed on staff both in supporting the technologies present at the shelter and in developing and maintaining expertise to effectively use those technologies in the provision of social services (Le Dantec and Edwards 2008b). Beyond the HMIS, whose presence was ubiquitous and mandated, the technologies present included mundane things like email, calendars, and shared documents, as well as a modest computer lab with a dozen personal computers for residents and their children to use for employment readiness training or with tutors who helped with homework and study skills. The presence of these technologies added complexity to the delivery of services, particularly managing shared calendars and documents in systems that were unfamiliar and very complex for the nontechnical staff at the shelter. The specific challenge was that work was continually re-created and left undocumented; shared folders and drives were understood to exist, but never integrated into work routines; and more tellingly, as has shown up in other contexts (Voida, Harmon, and Al-Ani 2012; Voida, Harmon, and Al-Ani 2011), the work to manage the technology was orthogonal to the work of providing social services, and so staff and counselors with limited time prioritized working with their clients rather than becoming system administrators.

These issues often meant that existing systems went underutilized because they added complexity to care provision. Enterprise-ready (even small-enterprise-ready) infrastructures require attention in initial setup and ongoing care to adjust and maintain so that they continue to match the needs of the organization. Because this kind of ongoing support was not present, the added overhead of new technologies made the staff hesitant to adopt new systems. Not unreasonably, they wanted support systems that required little management overhead and that helped them build and maintain the connections they needed to do their job: connections within the shelter to help share information and context with each other, connections to external organizations with which they exchanged referrals, and support to complement the information they were entering in the HMIS, especially within a climate where documenting outcomes was becoming a priority.

Second, just as the care providers had to manage relationships at an institutional level, they also had to develop and manage multiple relationships within the local context of the shelter. This issue had direct bearing on the residents of the shelter as well, because effective care often depended on the staff developing and maintaining close relationships with the mothers at the shelter. During the two years I worked with staff and residents at the shelter, there were three primary staff positions. The program director ran the shelter and was the main point of contact for developing programs of counseling, connecting residents to external programs and longer-term housing support, and establishing a routine and a set of rules to help impose order and stability for families arriving in crisis; two weekend caseworkers took on similar responsibilities during the weekends and complemented the program director's role as a trusted confidant at the shelter; and a night manager did not have specific counseling or support responsibilities but added another point of contact for the mothers. The personnel filling these positions were generally stable, with the exception of the night manager position, which saw steady turnover and meant new faces every few weeks.

The reality of social services is that the person-to-person relationship between provider and client is paramount (Le Dantec and Edwards 2008a; Hersberger 2001). As a result, the shelter staff were the residents' preferred source for information about social services and aid programs. This often led to situations in which the staff created a bottleneck in terms of helping each of their clients find resources. This bottleneck was most obvious for the program director, who was the primary point of contact for referrals and connections to programs that required her involvement with external providers to ensure her residents were being enrolled in needed programs. The issue was largely one of communication and limited time: she had new information every day that needed to be sent out to the mothers at the shelter, and the mothers had new information about their situations that they needed to ensure the program director was apprised of.

The third issue stemmed from the fact that care provision relied on collective action, requiring varying degrees of coordination between individual staff members as well as across distinct organizations. There was a mix of consequences the staff had to manage, some in relation to external accountabilities (as I developed in the previous section, the shelter staff had to account for services provided through the HMIS and generate reports for regulatory and funding compliance). The coordination work here was based on ensuring that services were first provided in a way consistent with the shelter's obligations to regulators and funders, and to accurately document the provision of those services in a way that was consistent across the different shelter staff involved in providing care.

Of much more daily importance to the shelter's operation was coordination between the program director and the caseworkers. During the week, the program director worked with the mothers to develop goals and plans of action. She had steady contact and provided additional help and encouragement as the mothers eased out of their immediate crisis and worked toward longer-term living and work situations. The program director would need to hand off these plans and the current progress to the caseworkers at different times, and likewise the caseworkers would need to hand off back to the program director. The high-touch nature of the crisis intervention programs being run at the shelter meant that these handoffs were complex moments of coordination in which the hourly details mattered: achieving different goals often required external service providers, and if a breakdown occurred, it was important to understand where and how it had happened so that the caseworker or program director could help correct it. And it mattered for the residents because their access to ongoing support (material and emotional) was tied to whether or not they appeared to be making good-faith efforts to follow through with their plan of action.

Woven through each of these issues was the staff's principled commitment to helping and supporting the families that came to the shelter. Their commitment was grounded in shared philosophies of care provision and social service, expressed through the service provider's mission, and which further defined the kinds of actions staff took in response to identifying and managing consequences facing their homeless clients. As with most organizations, tensions arose around how to act on those shared beliefs as they moved between individual motivation and organizational principle, but the common ground provided a solid basis from which to interpret and relate to the issues of resource and relationship management and the coordination work necessary to provide services and assistance to the mothers at the shelter.

Shelter Residents

The prima facie social condition confronting the women at the shelter was the fact of their homelessness. A number of more specific issues became manifest through the months of ethnographic work and the participatory design intervention. Like the staff, the residents needed to develop and maintain new relationships and needed to gain access to information, but they also needed to develop the capacity to act on that information as they emerged from the acute circumstances that led to their arrival at the shelter. The mothers also needed to develop their own sense of capability, to become empowered over the circumstances in their life so that they were not simply being swept along by events but would begin to feel that they could make decisions that would change their outcomes.

Developing and managing relationships had two components for the mothers at the shelter. First, they had to cultivate practices that enabled them to maintain existing social support. The loss of social ties, and the impact this has on the homeless, are well documented (e.g., Hersberger 2003; Le Dantec and Edwards 2008a), but the issue is not just the preservation of communication with friends and family members but preserving or reestablishing the moral and cultural ties that are imbued in belonging to a community (McMillan and Chavis 1996). The second challenge came from coping with the stigma of being homeless and the desire to maintain an image of stability for friends and family who might otherwise be unaware of their situation (Le Dantec and Edwards 2008a). Mothers in the shelter were contending with both issues; they often discussed missing the emotional support and companionship of former neighbors or friends, as well as missing the instrumental support such social networks provide for practical matters like looking after children and helping them juggle the multitude of responsibilities of poverty and single parenting.

The other component of this issue comprised developing and maintaining trusted relationships with the shelter staff and with fellow residents. The challenge here was that for many of the mothers, the default position with respect to social institutions and individuals offering help was one of distrust—either from previous bad experiences or as a result of going through personal upheaval. Building relationships at the shelter was crucial for successfully constructing and navigating a plan to return to greater stability and self-sufficiency: partly due to the resources gated by relationships to shelter staff and other providers, and partly in order to alter the socialization that may have been, in fact, inhibiting the mothers from gaining access to services and needed resources (Conley 1996; Snow and Anderson 1987). From the perspective of programs and services, the staff could only develop appropriate courses of action when they sufficiently understood the individual details of what was going on with the mothers; however, there was a deep tension for the mothers in being completely forthright, as many of those details were sources of, or contributed to, social stigma. The staff worked hard to make the shelter a safe place, physically and emotionally, but the mothers had to make efforts to engage with the staff. Once trust was established, then instrumental problems such as connecting to appropriate services became much easier. Just as important as getting connected to services, building trusted relationships with the staff extended the mother's social support network and created ties that would remain active and supporting even after they left the care of the emergency shelter.

Concurrent to the need to develop relationships with the staff was the need to cultivate strategies for contending with information overload. As Hersberger has pointed out (2001; 2002; 2005), the homeless are confronted with a plethora of information from different service providers; making sense of that information, rather than getting access to it, is the main challenge confronting them. Information overload was a substantial challenge for the mothers at the shelter. When they arrived, the program director would provide and step through a large folder full of programs and services the mothers would then need to familiarize themselves with so they could begin to co-construct their path out of the shelter. Everyone involved understood that there was too much information at once in the folder, but it was viewed as a starting point the mothers could refer back to and develop familiarity with over their time at the shelter. In addition to the information packet from the program director, the mothers would receive new details every day regarding programs they might qualify for or in which they should be enrolling; fellow residents would share information, particularly about potential jobs and apartments, but also about how to navigate the different services in which they were enrolled; and shelter volunteers were also providing information and suggestions about opportunities for the mothers or their children.

Plotting a course through all of these sources of information and the details that needed learning and acting upon was an enormous challenge for the mothers. As a result, the context in which the information was situated was often more important than the information itself (Alexander et al. 2005), which puts into further relief the importance of building trusted relationships with the staff and other residents at the shelter: those relationships became a filter and validator for information the mothers needed in order to begin rebuilding. For the shelter residents who developed close ties with the staff and their fellow residents, ordering and navigating the information was much more easily managed once the task became a cooperative and social endeavor rather than an individual challenge.

Running through these intertwined issues of building relationships and cultivating information-seeking practices was a more fundamental issue of developing a sense of empowerment and individual capacity while the mothers were at the shelter. One mechanism for empowering the mothers was through the programming at the shelter: job training, enrollment in financial aid programs, connection to health care services, and support for the young children each of the mothers was caring for. By relieving the acute housing crisis that led the mothers to the shelter, and by helping order the many individual (but related) challenges the mothers were facing, the shelter's programming helped relieve the overwhelming stress and empowered the mothers to tackle problems step by step.

Beyond providing services to the mothers, however, empowerment was at root a developing and enacting of agency in their lives. Empowerment as an issue at the shelter came through in how the mothers moved from close care and coaching from the staff to sharing and coaching each other as they progressed toward moving out of the shelter. And here rested one of the main opportunities in the developing design discourse at the shelter: cultivating and sharing expertise the mothers embodied as they came to enact more agency in their own situations. Where the shelter staff presented a particular perspective on the multiple programs and resources available to the mothers, the mothers' perspectives were as different as the recipients of those programs and services. Articulating their experiences and expertise through the issues of relationships and information practices, and treating them as valid and on equal standing with the knowledge and expertise from the shelter staff, became a vehicle for empowering the mothers, for elevating them from victim to source of support, and for creating an opportunity to enact agency in support of common goals and shared issues (Le Dantec et al. 2011; Le Dantec 2012). This empowerment was derived from the co-design experience by making their experiences visible as resources for design rather than symptoms of problems to minimize. Much in the way McCarthy and Wright describe the goal of participatory design projects as including a plurality of voices and experience (McCarthy and Wright 2015, 26), the mother's homelessness was a domain of expertise from which they could contribute insight as equal members of a design process.

Bounding a Public

Connecting the issues across the institutional and local scales begins to develop a more complete picture of the social conditions present within the context of social service provision for the urban homeless. At both ends, specific issues existed that were largely invisible from the far side: the regulatory and funding environment in which care providers must work is far removed and invisible to those receiving care; the specific circumstances and situations that contributed to any one family's homelessness are invisible to policymakers. In the middle, however, there is a specific site where people are enacting responses to these distant and distinct sets of issues. This site is where, as Dewey points out, we may find a public: indirectly and seriously affected by a set of social conditions.

As the different issues moved into currency throughout the larger network, they were transformed from local matters of fact into matters of concern: the specific and concrete problems being confronted at each node in the network—homeless families, caseworkers, social service organizations, regional and state regulators—became a complex of issues with unique affordances for each of those affected. These issues were circulated through the participatory design process—the service provider workshop, engagement with the shelter staff and residents—and through the artifacts of that process—the catalog of goals, resources, and information flows and the maps, sketches, and prototypes that led up to a deployable system. As the issues moved through this ecosystem, they were enacted in particular ways via the composition of actors, artifacts, and institutions involved. This puts issues at a confluence of theoretical perspectives in which Law, Latour, and Dorst intersect in the complexity of contextualizing issues through their multi-sited enactments (Law 2004), their contestability when moved between parallel local settings (Latour 2008), and ultimately the implications of these features as a framing for design where the wickedness of a design problem results from how that problem is constructed and enacted rather than being an inherent characteristic of the problem itself (Dorst 2006).

What occurred through the course of the ethnographic fieldwork and initial design workshops was a process of articulating the most proximal issues and situating those against issues more distant that were enacted at different scales and at different sites. The maps and documentation that were generated through the workshops became a shared resource that linked the different sites together. I then used those resources to work with the staff and mothers at the shelter to contextualize the daily issues they were facing as matters of concern, connected through the larger machinery of social service provision and social conditions.

The particular entanglements of issues between the staff and mothers at the shelter emerged through the early design process: for the staff, this meant focusing on ways to help cope with resource constraints, manage multiple relationships, and provide support for collective action as staff worked to support several residents at a time; for the mothers at the shelter, it included features to mitigate information overload, enable continuity in maintaining social support and trusted relationships, and finally the co-design of features to support personal empowerment through information sharing and the increased autonomy such practices enabled. These issues, and the specific ways they were bound up between the shelter staff and the residents, individuals and institutions, and artifacts and structures created material connections between two groups of stakeholders, or what might more aptly be referred to as implicated actors.

And this is precisely the point that Marres draws out with respect to the utility of issues and issue-bound publics (Marres 2007): a focus on issues is a way to break from the established positions of stakeholder groups where those involved are known and are invested in working toward resolution. Taking this familiar path to understanding the different interests in play, when we treat the policymakers, care providers, and homeless as separate stakeholders—each with particular needs and goals that need to be mediated or assuaged through design—we fall into reinforcing the differences between them. Moreover, the stakeholder position fixes the subject relationships of each group with respect to each other—through mechanisms of authority, access to resources, and forms of social and material production and consumption—rather than developing a more fluid subject relationship set against the common external social conditions with which they are each contending.

This critique of the term *stakeholder* for "individuals affected by particular conditions" is part of what sits behind the move toward the term *implicated actor*. In her formulation of the latter, Clarke (2005) draws attention to power relations and points out two particular categories of implicated actors: those who are physically present but marginalized and silenced by those in power, and those who are not physically present but who are conceived of as the targets of others' work (ibid., 46). The homeless mothers clearly moved between these two poles as implicated actors. Their relationship to social services was clearly defined as being the target of others' work and living with the stereotypes and stigmas of homelessness; and their voices were often silenced, their experiences rendered invisible by the systems of social care and the bureaucratic obligations of poverty.

Issue formation through design is a tactic in a larger strategy of social design that helps counter the marginalization of implicated actors. As pointed out above, and documented in more detail previously (Le Dantec et al. 2010; Le Dantec et al. 2011; Le Dantec 2012), when the mothers at the shelter worked through the design process, their experiences with the conditions of homelessness became a source of expertise rather than a target of others' work; their shared knowledge of moving through different systems of care became a resource for collectively resisting institutionalized power relations; and the shared visibility of work and social conditions between the staff and residents changed both the work and the social conditions at the shelter.

Operationally, constituting a public through participatory design means developing an understanding of the issues such that it becomes possible to erode the stronghold of existing subjectivities and create a shared space for further design investigations. The distinct sets of instrumental needs and goals that might otherwise be balanced against each other, or mediated through design, become a joint set of issues, connecting and motivating the further production of potential and proposed solutions. For the case presented here, the shelter staff and residents began to materialize as a public with many of these elements: as distinct groups with distinct goals, accountabilities, and modes of productive participation, their needs could naturally be decoupled; the design intervention, however, made the overarching social condition newly visible and created a shared ontological space for engaging with the complex ways in which shelter staff and residents were entangled with each other.

The shelter staff and residents needed to manage both different kinds of data and different kinds of relationships to those data. In some cases these were data from distinct sources with distinct purposes, but more often the data being managed were coproduced through the development of counseling programs and the work done to complete those programs. It was the interaction and feedback loops of the shelter staff collecting and sharing relevant resources, and the residents learning and making use of those resources, that drove both the production of data in the HMIS and the production of service relations between the staff and mothers. So while the staff contended with the issues of *coordinating action* and *managing resources*, those issues were bound to the residents' issue of *information access*, both of which were experienced through the issue of *managing relationships*. Likewise, the issue of *empowerment* for the residents was experienced in relation to *information access, coordinating action*, and *relationship management* as programs were constructed to emotionally and materially build the mothers up so they could move on from the shelter.

A focus on issues as constitutive of a public also becomes a means of navigating across networks and the kinds of specific concerns that are imposed by different kinds of accountabilities. The institutional issues of data and relationship management were enacted in the specific ways the staff interacted with the mothers at the shelter. The kinds of counseling and skills programs staff directed the residents toward, and the ways they structured their interactions so that progress could be documented and strategies adjusted, were each done with consideration of how those activities would then be managed into the HMIS and how and whether they were connecting to other service providers within their trusted cohort.

The most important point here is that each of the issues, whether located with the shelter staff or with the mothers, was experienced as a relation to issues in the other group. At times these issues were exacerbated by the other, as when the caseworker or program director did not coordinate and as a result bad information was provided to the resident, or when the resident did not follow the rules at the shelter, creating an adversarial atmosphere in which shelter staff had to occupy the role of authoritarian rather than confidant. That the shelter staff and residents have complex and contentious relationships does not weaken their participation within a public; it is, in fact, the contrary and complicated nature of their interaction that makes a public a useful context in which to understand how both groups are bound together through a set of shared issues. Negotiating shared responses to those issues and building dynamic constituencies so that some collective progress can be made is what makes a public important for social design. By shifting away from simply considering the siloed needs of each group on its own, constituting a public through a recognition of shared and entangled issues delineated a design space in which the care providers and homeless mothers were not treated separately as producers and consumers of information or services, but as a collective body in which the condition of homelessness must be contended with together, even as distinct accountabilities must be addressed and satisfied.

4   Identifying Attachments

In the previous chapter I outlined the ways issues were articulated through the design process and how those issues, once reflected upon and integrated into the developing design discourse, created a basis on which to begin constituting a public. Seeking to address present issues is not, in and of itself, new to design; however, the focus on issues here is meant to express a network of actors, artifacts, and institutions in a community setting, rather than define product features. When we examine the relationships at the shelter as co-constructed and entangled in contending with the social conditions of homelessness, the barriers between implicated actors begin to break down: the shelter staff are no longer only producers of social services, and the shelter residents are no longer only consumers of those services. Instead, the staff and residents are connected through a set of issues which they are taking distinct but complementary and coordinated action to resolve. The mere presence of shared issues, however, is not sufficient for the formation of a public. There needs to be a shared set of commitments and dependencies that connects distinct actors, artifacts, and institutions. Here I return to the concept of *attachments* to more fully develop how they operate as the bonding agent, cohering the plural experiences of shared conditions into a public.

A useful way to begin to conceptualize attachments is through their dual role as "commitments to" and "dependencies on," each of which creates an obligation and a reliance on the relation between actor, artifact, and institution (Marres 2007). The mechanics of that relation depend on the development and exposition of affective interactions between constituents, which in turn set the stage for cultivating shared values that cement those attachments. Affect, in the sense that I will use here, differs from its legacies in artificial intelligence and human-computer interaction, which focus on the ways computing experiences can structure and represent emotion (Picard 1997), or how emotional responses can be used as a resource for designing interactive experiences (Sengers et al. 2008). Instead, I am turning to scholarship in the social sciences and critical theory rooted in the Spinozan principle of *affect* that signifies "the augmentation or diminution of a body's capacity to act, to engage, to connect" (Clough and Halley 2007, 2).

Affect, in this sense, relies on an understanding of affective experiences as separate and prior to emotions. This marks a critical departure from the legacy of affective computing, in which affect and emotion were synonymous and the goal was to encode emotional experiences as units of information on which to operate (Picard 1997). To illustrate the difference between an affective experience and an emotional experience, I can provide a brief scene: if I were to suddenly and violently yell at you for no reason, you might flinch and experience a bodily response—an affective response. However, if I immediately yelled at you again, you might feel anger or confusion or annoyance at being yelled at—an emotional response.[1] In the first instance, your bodily reaction occurs prior to encoding an understanding of that response; in the second, you encode your response as a particular kind of emotional experience. This distinction is central to an understanding of affect as "a prepersonal intensity corresponding to the passage from one experiential state of the body to another and implying an augmentation or diminution of that body's capacity to act" (Deleuze and Guattari 1987, xvi), or as Gilbert (2013, 146) says more plainly, "[Affect is a relation that] enables humans to share the feelings of others, and to have their own affective states altered by events which occur in the lives of others, without having any rational, personal or self-interested reason to do so."

Recent theorists have demonstrated, across a range of contexts, the ways that affect provides not just another vantage from which to theorize the social but the tools to account for sociotechnical relations, drawing attention to the ways these relations augment different capacities to act (Clough and Halley 2007). What this vantage affords us here is a way to understand how actors influence and constitute capacities for action. These capacities create and arise out of a shared ontological space in which actors produce and are moved by affective interactions (Grossberg 2010, 311). For Grossberg, however, the understanding of affective interactions as constitutive of a shared ontological space is not the key concern of analysis. Rather, he is concerned with *how* that ontological space is articulated in empirical experiences through "regimes of discourse that [constitute] the ways in which we live our lives" (ibid., 314). Attachments inhabit the flows between the *ontological* affect and the *empirical* experience that accompany the formation of a public. Further, while affect develops a capacity to act, when enacted through design as a machinic assemblage (Deleuze and Guattari 1987), affective connections facilitate the passage from ontological possibility to empirical reality by way of attachments and the generative moves to address particular issues.

Without attachments, and the motivating affective interactions, the formation of a public as an entity capable of co-constructing material responses to shared issues fails. Attachments are critical as they build out the collective capacities to act on issues. In the empirical setting at the center of this book, the development and articulation of attachments determined the degree to which a public emerged out of the practices at the shelter. In some instances, new attachments enabled the staff and residents to approach issues collectively and to develop new practices in the process. In other instances, articulated attachments exposed the authority and power dynamics at the shelter and created opportunities for those to be reconfigured. In still others, the attachments that became manifest did not constructively motivate action, but did serve to underscore the role of affect in that outcomes either augmented or diminished capacities to act in specific ways. Throughout, the staff and residents at the shelter worked to design community technologies that would support their practices and augment their collective ability to act.

The issues that were articulated through the early participatory design process became a basis for understanding the attachments between the shelter staff and residents. It is at this point that a shift in focus begins to take place in the empirical analysis. The previous chapter focused on issues and the ways in which different social and technical configurations created them and influenced how those issues were made manifest and how they circulated as matters of concern; here, by tracing the design process from the early stages to a period of appropriation and redesign, the focus falls on the way activity at the shelter was configured and reconfigured around the new technical capacities made available through the co-designed technology. I would point out that the conceptualization of issues among the staff and residents also matured during this period. The consequences of focusing on information sharing and coordinating action, and of committing to particular technical responses to support relationship management, each brought into greater relief the ways these issues shaped and were shaped by the relations at the shelter. During this time, the point of interest shifted to the staff's and residents' changing relationships to those issues as the nascent design practice and new technical capacities mediated their experience of them. The evolution and creation of attachments becomes easier to trace, made visible by the appropriation of a new set of technical capabilities, and highlights the sociotechnical interplay as actors, artifacts, and institutions are enmeshed into a public contending with proximal and distant social conditions.

Before continuing, it will be useful to briefly outline the new technical capabilities that were co-designed with shelter staff and residents. The system included three components: staff at the shelter accessed the system through a web application called the Message Center, which provided an interface for sending and receiving short message service (SMS) messages to residents' phones; shelter residents interacted directly with their mobile phones, sending and receiving messages from the staff throughout their stay at the shelter; both staff and residents had access to a mounted display in the shelter called the Shared Message Board, where each could post information and announcements either via SMS or through the Message Center (see figure 4.1). Throughout the project, I had the good fortune to collaborate with Wendy Kellogg's group at IBM Research to implement and deploy the system. The development team, including Mark Bailey, Jim Christensen, and Rob Farrell, drew on their previous experience building systems to share information via mobile phones, and we were able to quickly produce and deploy a robust prototype able to withstand daily use at the shelter.

The goal of the initial feature set was to support the sharing of information and resources at the shelter, both between staff and residents and just as importantly among the residents themselves. For the staff, this support largely turned on providing tools that helped them amplify their contact with each resident via the Message Center and the use of SMS messages. These tools were an important advance within the shelter that allowed the staff and residents to communicate asynchronously and at times that would not otherwise have been possible (Le Dantec et al. 2011). To support the residents, unmodified mobile phones were selected as the primary interface due to their familiarity and prevalence, even among the very poor and

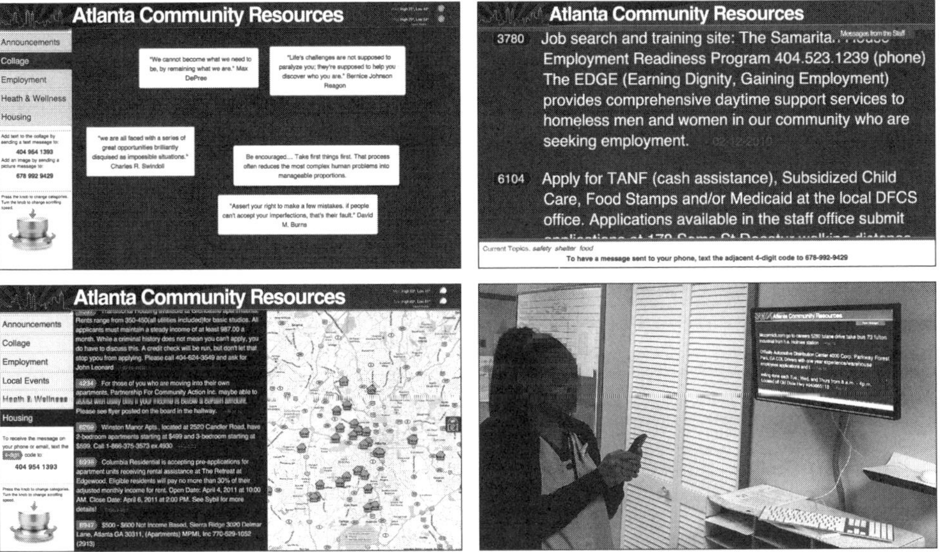

Figure 4.1
Shared Message Board screenshots and placement within the shelter.

homeless (Le Dantec and Edwards 2008a; Woelfer et al. 2011; Rice, Lee, and Taitt 2011).[2] The Shared Message Board was designed as a common space where both staff and residents could post messages, and where messages would all have equal visibility regardless of provenance, thus providing a platform for the residents to articulate issues in a setting that presented their knowledge and experience with equal visibility to that of information provided by the staff.

Through these new technical capabilities—a web application that enabled asynchronous communication through SMS, and a shared digital message board in which institutional and tacit knowledge were equally commingled—attachments were enacted and formed anew. Bifurcating these attachments as a set of "commitments to" and "dependencies on" that were enacted by those at the shelter enables us to develop a more robust understanding of how the design interventions developed parallel capacities to act among co-designers and through the practices of reestablishing stability via the programming at the shelter. Many of the conditions for affective relations already existed at the shelter—stemming from genuine dedication to helping the families at the shelter regain their footing—but as the design process matured and the technology was deployed, appropriated, and iterated upon, the relationships to issues evolved, prompting new attachments to form and altering how existing attachments were enacted.

Commitments To

During the design of the system, much of the discussion was grounded in the everyday constraints of current work practices and information needs at the shelter: the relationship between shelter staff and residents, the accountabilities and obligations of shelter staff within their regulatory context, the differentials in responsibility and institutional influence between residents and staff, and the need for communal support among the residents. These acute arrangements shaped how the staff and residents engaged in the design process and narrowed the possible activities that might be mediated by new technical capabilities—focusing on the forms of communication, the kinds of information shared, and the means of making visible different perspectives on shelter life. Through these practical issues, the staff and residents began to articulate their commitments to the social outcomes meant to help the mothers, and the technical capabilities they were designing and appropriating to achieve them.

Staff: Acting on Established Commitments

The shelter staff were committed to helping the residents and to providing programmatic support and individual encouragement. Above all, the staff strove to create an environment in which the mothers could develop a sense of empowerment: to begin to plan for the future—however near-term that might be—and to take ownership of the decisions they were making. This commitment shaped the programming at the shelter and the interactions between the staff and the residents. For the program director, empowerment was instrumental and was tied to gaining control over finances, finding secure housing, and situating children in school or day care.[3] She laid down the rules and the procedures of the shelter and then expected the mothers to follow those rules and participate in the procedures so that in several weeks' time they would have built a sense of capability and empowerment and could move on from the emergency support provided by this shelter. Typically, this meant finding a delicate balance between a job that paid well enough, securing affordable housing close to the job, ensuring reliable transportation, and enrolling their children in school or other forms of care. The procedures and instrumental approach the program director used helped order each of these individually daunting and collectively paralyzing challenges: first find work, then find housing, then enroll the children in school. For the program director, the commitment to empowerment came through helping the mothers feel that their problems were tractable, that they could manage them, and that the procedures she used had worked for the many mothers and families who had come through the shelter previously.

As a foil to the program director, the caseworker's commitment to empowerment was expressed more personally through developing rapport with the mothers as their confidant and trusted insider: she positioned herself in a way to support the emotional well-being of the mothers in order to complement the instrumental focus of the program director. Because the caseworker spent most of her time with the families over the weekends—while the program director was the primary contact during the week—she was able to work with the mothers during times that were less driven by the demands of the workweek. The result was often a set of close relationships with the mothers in which the caseworker was less overtly focused on the weekly progress of the mothers in their individual programs of care and more attentive to where gaps were appearing in those programs. By cultivating closer relationships with the residents, she built cooperative opportunities in which the structure and priorities of what the residents were working on were co-created rather than enforced didactically.

Together, the program director and caseworker created a coordinated team in which the authority of the program director worked to intervene in crises and to structure programs to create new routines, while the caseworker helped build confidence and independence, supporting and encouraging the mothers as they worked to reestablish themselves. All of this work relied on close collaboration and sharing between the program director and caseworker: both needed to know how the mothers were faring during their stay at the shelter, and there were two clear points of transition each week, on Friday and Monday, when progress or any particular issues to be attended to were communicated.

As described in the previous chapter, the need to coordinate action, and to create continuity over the course of any given week, was a substantial concern for the staff during the design of features in the Message

Center—the web application the staff used to communicate via SMS messages with the residents. Initially, the Message Center was conceived as a way to provide just-in-time information to the residents and to open an additional channel of communication between residents and staff—one that was not bound to the limited hours staff and residents were together in the evenings.[4] Coordination activities were viewed in light of staff and residents sharing information throughout the day and both being able to adjust to circumstances, particularly around enrolling in external services and programs, for which the program director or caseworker would often need to connect to staff at the external organization as part of the referral process (all of this tracking the institutional issues discussed in the previous chapter). However, it was not until early prototypes of the system began to circulate that the staff saw another opportunity for using the Message Center as a point of coordination: opening up the Message Center so that each staff person could view all messages, regardless of who sent them or to whom they were sent (Le Dantec et al. 2010). This change in direction was substantial, as the initial prototype treated messages much like email while the altered design was more akin to an Internet forum. Initially, the staff person would log into the system and see the messages sent to them by each resident, along with a record of messages they sent to each resident. Now, staff would first see messages sent directly to them, but they could also view all messages transacted by the Message Center, peering into conversations they had not directly been a part of.

Motivating this shift was the staff's need to "keep the pulse" of the current topics and problems the residents were focused on in order to help the weekday-to-weekend transitions. Once the system was deployed, the ability of the program director and caseworker to view each other's digital conversations with the residents began to shift the way particular attachments were articulated and the way each enacted their commitments to empowering the mothers at the shelter. The most significant change grew out of the way the division of labor was refactored. The caseworker voluntarily took on most of the responsibility for updating information to the Shared Message Board—the public display in the shelter—and using the Message Center to communicate directly with the mothers. As a result of taking up the technology to further communicate with the mothers, the balance of power shifted: by virtue of using the Message Center to communicate with residents and post information to the Shared Message Board, the caseworker's role at the shelter was made more visible and amplified. She was no longer only present during the weekends, but was now able to offer guidance and support to the mothers throughout the week. Furthermore, because her role within the practice of shelter life had developed out of a desire to build rapport and supportive relationships with the mothers, the way she used the Message Center to communicate with the mothers was by engaging in conversation and collaboration, not simply directing them to information they needed to act on (for a much more detailed account, see Le Dantec et al. 2011). The consequence was that dormant differences between the program director and the caseworker in their commitments to empowerment and how to most effectively intervene in the mothers' lives were made visible and amplified.

While the move to make all messages available to the staff was done to enable better awareness of the particular issues the residents were working on, that awareness was limited and imperfect. The brief nature of messages sent to mobile phones (typically adhering to the 160-character limit of a single SMS message) meant that the larger conversation, one that often started during a face-to-face meeting, was difficult to follow. Furthermore, the forum-like qualities of the Message Center became an invitation for the program director to surveil the activities of the caseworker. In one instance, the program director reproached the caseworker for contacting a former resident with what had been mistakenly interpreted as an invitation to an event for current residents. The challenge here was not specifically one of exercising authority over subordinate employees, but of enacting commitments in different ways. The program director was very focused on supporting the mothers who were currently at the shelter, as limited resources meant she needed to ensure they were directed at those immediately under her care. The caseworker saw the technology as a way to extend her relationships to the mothers, whether currently at the shelter or not, and the efforts to continue to share information with current and recent residents was the same—following similar contours to the use of email in a time when it has made the cost of widespread communication effectively zero (Shirky 2008). Moreover, because maintaining relationships with former residents was more readily mediated by the Message Center, the caseworker could continue to offer cooperative support as she had while the mothers were at the shelter.

As use of the Message Center developed, the program director and caseworker had to negotiate what was and was not appropriate use of the technology, establish boundaries that had not previously been necessary, and come to an agreement on how to productively integrate the new capabilities provided by the Message Center, given the different roles they had traditionally occupied within the shelter: one that was instrumentally and operationally focused on running programs to help the mothers rebuild their lives, as embodied by the program director, and one that grew out of providing social and emotional support to the mothers as tools to help manage the overwhelming conditions of homelessness, as embodied by the caseworker. What shifted were the attachments to a set of issues that had always been present but were now being contended with differently. The issue of maintaining relationships with the residents was always present at the shelter, and even the need to maintain longer-term relationships to enable follow-up as a means to arrive at outcome-based programs. The technology, though, prompted a reconfiguration in the attachments to those issues as new capabilities became available and as those new capabilities altered how and when the caseworker in particular could act on her commitments to the mothers.

Once the caseworker's visibility and availability shifted via her technology-mediated contact with the mothers, she became the more readily reached point of contact. This was important beyond the simple instrumental consequence of enabling the mothers readier access to information, because the role the caseworker occupied at the shelter was one of providing affective support: she facilitated the creation of an intimate environment of home, offering support and encouragement, and was more overt in approaching the mothers in partnership, working jointly to help them succeed in the different programs in which they were involved (job training, family counseling, etc.). Her commitment to the mothers was on an individual basis, instead of a programmatic basis as the director's commitment was, and the Message Center, as a tool of personal (if not private) communication, amplified her ability to establish those interpersonal attachments, creating a cooperative environment in which overcoming the conditions that had led the mothers to the shelter was framed as a set of issues to be confronted together.

This collective sense of shared problem solving arose out of the new channels of affective labor the Message Centered enabled within the shelter—such as the late-night messages and words of encouragement and feelings of connection I have documented in greater detail elsewhere (Le

Dantec et al. 2010; Le Dantec et al. 2011; Le Dantec 2012). As patterns of work shifted around these channels, attachments were enacted through a set of differentiated commitments. Some of those commitments grew out of familiar work relations (staff were paid to care for others), while other commitments were rooted in circumstance (the mothers were obliged to attend to programs of service). Finally, there were a set of interpersonal commitments as particular staff and particular mothers grew to care for each other outside the coercive regimes of employment and circumstance. Throughout, a design process mediated these changes.

Instead of focusing purely on the instrumental relationship between staff and residents, one aligned with the provision of services that treated the staff as producers and the residents as consumers, the design process that begat the Message Center enabled the staff and residents to enact new commitments: commitments grounded in developing cooperative rather than service-oriented relationships; commitments grounded in the affective labor that enabled connections beyond the limited in-person time. These commitments were tied to a kind of intimacy both staff and residents wanted to cultivate at the shelter. Even as the shelter was a place of work for some, and while other work circumstances have pointed to the ways such always-on connections introduce a collection of negative consequences for those involved (Gregg 2011), the shelter was also a home, and so the affective work that grew out of intimacy had an opposite effect—rather than creating precarity in planning daily activities or responding to mundane issues, it established a stable network of social support. However, the tension between workplace for staff and home for residents meant that the shared commitments to creating a familial environment muddled the dynamics of the workplace into a kind of common cause where being available long after hours was not just the expectation of a (newly) high-tech workplace, but a commitment to directly impacting families in need of help.

The new capacities to act rippled through the kind of activities the staff and residents undertook. The caseworker voluntarily took on additional responsibilities by communicating with the mothers via the Message Center. This often meant doing so well beyond the hours she would normally be present at the shelter—from home and from her other place of employment. Her willingness to do this was tied to an intentional feature of the technology. Unlike the mothers, who would receive messages directly to their cell phones, the Message Center did not forward messages directly to the caseworker's cell phone. She had had previous experiences in which giving out her personal number created long-term challenges, as current and former residents would sometimes abuse that direct connection. The Message Center, however, gave the impression of a direct connection while still enabling the caseworker to choose when to check for and respond to messages from the residents. There was a clear imbalance of power here as the residents could not choose when they would receive messages, but the design choice was made to manage two different kinds of needs: the daily needs of the staff to be able to attend to their responsibilities and manage communication within those responsibilities, and the needs of the residents to have a ready link to support. Just as the staff adapted to the technology, the residents also worked under new expectations of when and how often they should seek assistance from the staff. When problem resolution was limited to in-person meetings, the schedule and routines under which they lived deferred some of the urgency in daily challenges. Once the mothers could ask for help via text messages, the staff could provide in-the-moment support which helped prevent small problems from derailing progress, but which also introduced a new nervous urgency throughout the workday for both staff and mothers.

Mothers: Developing New Commitments

While at the shelter, the residents' commitments to different issues followed an arc. Initially, as they arrived in the midst of crisis, they were singularly committed to stabilizing their situation and caring for the immediate needs of their families. Over the handful of weeks they spent at the shelter, the residents would develop commitments to each other and to the community that was present at the shelter—the staff as well as the stream of volunteers who arrived each night to prepare and serve meals to the families. Traversing this arc from crisis to experienced community member inculcated a sense of responsibility to—a commitment to—fellow residents and a desire to help the families newly arrived and disoriented. In turn, the process of co-design and of reflecting on the early prototypes was shaped by where the particular generation of mothers was along this arc. In the initial days of co-design, the participatory activities were focused primarily on building an understanding of design as an activity and practice they were capable of contributing to, of undermining my own authority as an outside expert, and drawing out their willingness to participate. The next step was to create a sense of empowerment so the mothers would engage critically with the design, relying on their experience as a valid perspective from which to critique design ideas. The final step was to arrive at a situation in which the residents had integrated the technology into their practices at the shelter, which articulated new commitments to each other, the staff, and the community.

The staff mostly encountered the technology through the Message Center in which they sent messages directly to the residents or added items to the Shared Message Board (see figure 4.1). For the residents, the messages received via the Message Center were viewed simply as messages from the staff and were not attributed specifically to technology we were co-designing;[5] the technology disappeared and the content being transmitted was the focal point. The most conspicuous encounter with the technology for the residents was with the Shared Message Board: it was a large screen placed at the entrance to the shelter, just above a table where the residents had to sign in every evening when they arrived. Locating the Shared Message Board so prominently was intentional, following guidance from the staff and residents, in order to draw attention to the information added to the board. While the initial goals of the Shared Message Board were to provide a visible place for both staff and residents to share information, as the design and deployment of the technology matured, the interface grew into a mechanism for surfacing common issues and enacting new attachments.

To illustrate this, a message was posted to the board during an incident when some residents were not completing their chores. The conflict had been simmering for some time, with several minor confrontations between the staff and one particular resident. It finally developed to a point at which another resident felt compelled to post a public message to the Shared Message Board:[6] "We came to [the shelter]. It was a blessing for us all and we knew the rules right away. ... We agreed to do them but we're not [all] doing [them]. This is the right thing and [we] count our blessings. ... Help to keep it clean." What is not captured by the text of the message alone is the atmosphere of support that the mothers were developing. By posting the message in a public way, the resident connected to the shared commitment to the shelter, a commitment to complete chores not simply as rules to be followed but as a commitment to the shelter (staff and building) developed from the "blessing" of being taken in during hard times. Further, the public nature of the message communicated that the commitment to the well-being of the shelter, and the practice of shelter life through the routines and programs, was integral to membership and a newly formed, if temporary, identity.

The commitments that developed were an important component of the formation of a public at the shelter. Mothers arrived as individuals with a broad set of common issues, but the distinct particulars were far more prominent. Over time, as they emerged from crisis and began to adopt the practices of shelter life, the mothers became residents at the shelter and the common and mundane issues became more visible, presenting themselves as potential points of attachment. The process of co-designing and adopting the Message Center and the Shared Message Board created opportunities to reflect on the shared issues being faced—from the common but individual issues bound up with their homelessness to the intimately shared but mundane issues of managing communal living in the shelter—and through that reflection, the staff and residents were able to imagine and enact different commitments to each other. A greater part of those commitments came via sharing information and shifting the accessibility of roles within the shelter; a smaller but no less important part came through enabling the residents to articulate their shared identity, their shared membership in the public that was emerging at the shelter.

Dependencies On

In concert with the commitments that developed over the course of the design and adoption of the Message Center and Shared Message Board, dependencies also materialized and were mediated through the creation and deployment of new social and technical practices at the shelter. These dependencies grew out of and relied on the commitments that took root among the staff and residents at the shelter, resulting in a tight feedback loop in which staff commitments to empowerment were linked to dependencies on the technology to organize and order information congruously with program structure, where resident commitments to an ephemeral shared identity depended on the shared social practices of contributing to shelter life. It was through exchanged commitments and dependencies that attachments at the shelter were completed, grounding the emergent public in both a set a common issues and a series of social and technical relations and practices in connection to those issues.

Staff: Dependencies on Practice

For the staff, the need to document different forms of interactions with residents was implicated through a number of different obligations and organizational goals. At the institutional scale, the staff needed documentation to facilitate compliance via records entered into the HMIS, and at the local scale, they needed documentation to support cooperative work and collaboration at the shelter. Importantly, that collaboration was not just among the staff—coordinating the actions of the program director and the caseworker—but was constructed around collective practices that required active participation from everyone involved, practices that were essential to the functioning of the shelter. In the most immediate sense, without active participation from the residents in the programs the shelter ran, the best outcome that could be hoped for was a momentary reprieve from long-term homelessness.[7] However, when residents were more engaged, they created an opportunity to turn the temporary relief of the shelter into more lasting changes in their lives. The staff aimed for these more lasting changes, and so when they delivered programs and developed relationships with the mothers at the shelter, they did so with commitment to empowering the mothers through education, training, and support systems for parenting so that after they left the shelter they would have their own capabilities to draw on. Through the commitment to empowerment, the staff also depended on the mothers to engage and be active members of the shelter community.

Even though the dependency on cooperative work practices had longer-reaching consequences for the residents, it was the staff's contact with the deployed technology that immediately reconfigured how they mediated their dependency on constructing a set of shared practices. Many of the changes to daily cooperative work practices were connected to the regulatory environment in which the staff provided services—the staff needed procedures and mechanisms for coordination and to document the programs and progress of the residents. The dependencies the staff operated under ran through a range of different practices, from informal tacit coordination around specific events—setting up meetings, following up on external referrals, sharing information about current counseling or care programs—up through formalized coordination practices for documenting and reporting services via the HMIS. The staff's work was an accounting of transactions, of interactions with external services, of interactions between staff and resident, or of interactions among the residents.

Taken together, the staff developed a complex dependency on the technology in which their perception of the technology as a way to convey information to the residents was transformed, from their initial design decisions to support information sharing to ways of using the technology as an explicit means for supporting individual empowerment. Likewise, information sharing practices that began as local and tacit coordination between the program director and the caseworker became repurposed and expanded into more systematic documentation practices that the staff came to depend on. In each case, the adoption and integration of the technology into shelter life evolved the work practices of the staff and interactions with the residents. New practices emerged as the staff and residents took ownership of the technology, integrating social and technical reconfigurations into how they approached shared issues and navigated common attachments to those issues.

Early on, the staff used the Message Center to send personalized messages to the residents. Some of these messages were shared across several residents as basic reminders or common needs, and issues were delivered and dealt with in the course of week-to-week life. As the caseworker became more comfortable with the technology, and as the residents began to respond to the messages with more regularity, she turned to heavier use of the technology as a way to extend her contact time with residents she did not see on a daily basis. Initially, her messages focused on the programs and services the mothers were involved with and concerned whether and to what degree they needed additional details, information about, or directions to those services. The caseworker's use of the Message Center quickly grew into more personal messages of encouragement and support (Le Dantec et al. 2011), and the SMS-based communication channel quickly became a default mechanism for contacting the mothers in the shelter.

An example of this arose during an early part of the deployment, once many of the more instrumental practices around using the Message Center and Shared Message Board had been established. As pointed out in the previous section, a conflict within the shelter had begun to develop around the completion of required chores. One of the consequences of this conflict was the residents articulating their commitment to the shelter. Another consequence was a demonstrated shift in how the staff, particularly the caseworker, strove to resolve the conflict. Prior to the deployment of the technology, conflicts were handled with face-to-face confrontations between the staff and the residents. After nearly two months of using the Message Center and Shared Message Board, the staff had begun to depend on the ability to send messages directly to the residents in place of some of that face-to-face interaction. During the conflict around completing chores, the caseworker used the Message Center to send a message to several residents, writing, "I am not sure whose turn it is but I have knocked on your door to remind you that the chore assigned to your room was not completed. Please take this opportunity to determine who needs to sweep and mop the kitchen. Thank you M—."

This particular message marked a shift toward managing confrontation via the Message Center, with the staff coming to depend on the technology to mediate certain kinds of communication. That dependency grew out of several features of the technology: the Message Center enabled the staff to communicate at times when face-to-face communication was not possible, it supported bulk messaging to all residents, and it created a record of the interaction that could be and was referred to in subsequent meetings. Each of these capabilities became integrated as features the staff depended on as they worked with the residents over the duration of their time at the shelter. Yet, more than these operational dependencies, the ability to send messages to the mothers was tied to cultivating responsibility. In the caseworker's words, "it returned power" to the residents, allowing them to choose how and when they would respond to a given message. By respecting the residents' boundaries, the caseworker returned agency to the residents, giving them opportunities to prioritize and respond to different issues. This was a subtle but important shift at the shelter, where prior practice had centered more on the staff setting priorities for the residents and guiding them through very structured programs. By moving some of the communication and relationship management practices to the technology, the staff could shift where the "work" took place in setting priorities, thereby empowering the residents to make decisions for themselves on issues both extraordinary and mundane.[8] This point is important on two counts. First, it further highlights the commitment to empowerment expressed by the staff, and how that commitment guided the caseworker's actions with the residents. She came to depend on the technology to mediate her relationships with the residents, extending her direct contact beyond the limited time she spent at the shelter. Second, it illustrates how enacting empowerment depended on the altered social boundaries afforded by the Message Center. The caseworker became more visible to the residents by virtue of her participation through the technology, and as her presence in their lives grew, it did so with a motive of delegation and empowerment in which she was much less concerned with micromanaging their lives and was more interested in providing opportunities and encouragement for the mothers to actively intervene in their own situations.

Concurrent with the communication practices that developed around the Message Center, the Shared Message Board also became a central channel for the staff to communicate to the residents. Once the staff and residents had become habituated to the use of the technology during the initial deployment period, two specific changes emerged as a result of those new work practices. These changes were incorporated into the ongoing design of the technology and into the social structures at the shelter. First, the staff wished to be able to present information on the Shared Message Board in a more categorized manner. Second, the residents wanted an easier way to find new information on the Shared Message Board. In both instances, the desire to alter the design turned on the utility of being able to post and find relevant information via the publicly mounted screen, and the challenge of navigating that information once it had become a regularly used channel for messages from the staff, posts from residents, and public exchanges between the two.

Throughout each of these practices, by sharing information through the newly created communication channels and by empowering the residents by respecting their agency in making choices about when and how to act on their responsibilities, the staff developed a much more critical dependency on the residents and their willingness to act. The shelter always demanded participation by the residents, but the new practices that developed in response to the design and deployment of the Shared Message Board and Message Center only strengthened those dependencies. Where previously residents had only to engage with programming at the shelter, once the staff shifted to using the technology to communicate with the residents, they had to engage with the technology as well, participating with people and computing artifacts in order to achieve the near-term goals established at the shelter. This dependency created an obligation for the residents to rise to, one that varied in character but which staff rationalized as contributing to empowerment by requiring residents to take more responsibility in acting on information and exercising choice and agency in the coproduction of regimes of counseling and social services. As these attachments took root, they created affective bonds between the people and the deployed technology, enabling new capacities to act through collective, sociotechnical participation in the issues facing the public at the shelter.

Mothers: Dependencies on Programs

The most explicit expression of dependencies from the residents was their immediate need for housing, meals, and services to help them connect to external programs for employment support, childcare, and legal aid (among others). During the early fieldwork and initial design activities, residents were dependent on the staff for guidance through the assortment of available and potential social services they might encounter and make use of. Such a dependency is to be expected and was only part of the mothers' experience when they arrived at the shelter.

Upon arrival, many of the mothers surrendered their needs to the shelter. The toll of events that led them to seek shelter was emotionally and physically exhausting, and in the first few days of residence, the mothers often simply let the staff and the routines at the shelter carry them. The staff's immediate job was to help the mothers regain emotional footing and a feeling of safety and care for their children so they could reclaim the dependency on the shelter and convert it into a dependency on themselves.

When the design work began for what would become the Message Center and Shared Message Board, the transition from the mothers' dependency on the staff to dependency on themselves was more explicit: the mothers most recently arrived at the shelter were given priority for the limited face-to-face time of the staff. Residents who had settled into their routines and were working toward specific goals during their remaining time at the shelter spent less direct time with the staff and, in particular, less direct time with the program director, who was most involved with new arrivals. In designing the features of the Shared Message Board, several of the experienced mothers discussed the desire to post information they had learned so that it would be accessible to future mothers via the screen: everything from specific details on programs or housing facilities they had good experience with to simple words of encouragement—helpful urgings to what they saw as their past selves, struggling to support their families.

It was in these design activities that the shift from a locus of dependency on the staff to a locus of dependency on the larger network of actors, artifacts, and institutions was most clearly enacted. Moreover, that shift in dependency appeared as a useful affordance for building up new practices of information sharing, creating affective connections to the staff and to the digital resources. The Message Center became the natural way for the staff to more regularly satisfy the residents' dependency on them during their early days at the shelter. By being able to communicate with the staff at times when they most needed help and direction, the residents were able to more successfully navigate the services and tasks they had to manage. The messaging between the staff and residents was largely made up of reminders and micro-coordination communication, such as directions to offices for different external services, connecting with staff at different agencies to verify enrollment in the shelter (many social services for the homeless require verification of homelessness, which is tied to demonstrating residence in a shelter), receiving reminders early in the morning to collect certain necessary documents for the day, and messages about new or short-lived opportunities for employment and housing.

This instrumental use of the Message Center to coordinate specific service procedures (e.g., securing long-term housing, following up on employment possibilities, or managing child care) was crucial to satisfying the residents' basic dependency on the staff. And enabling that dependency to be acted on in situ while the mothers were away from the shelter helped the process of accomplishing goals in the world. It became clear over time, however, that the most important mode of communication between the staff and residents was phatic—that is, messages whose content built and extended rapport between the staff and residents (e.g., "I really enjoyed the meeting yesterday evening. Perhaps I should have more bonding and sharing experiences," and "Thank you, I look forward to talking to you too"). The caseworker started sending such messages to the mothers as a way to gently encourage and connect with them while she was away from the shelter. Particularly since she was not present for most of the working week, she did not have up-to-date information on what they might be working on or what specific instrumental information needs they may have had. So instead, she focused on sending messages that supported the emotional well-being of the mothers, and in doing so, made what had been a tacit acknowledgment of the mothers' emotional dependency on the shelter more explicit; she acknowledged and respected their dependency through a series of affective interactions that built rapport and trust.

In instances when the staff and residents engaged in relationship-building messages, there was also an increase in instrumental messages. This positive feedback loop points to how the issue of relationship management shared by the staff and residents was mediated by a dependency on trust and feeling of common cause; as the staff and residents became copartici-pants in creating and managing relationships, they were able to work more actively toward resolving additional shared issues, dealing with the condi-tions of poverty that had led to the mothers' homelessness. Conversely, simply having increased access to the staff via the Message Center was help-ful, but not as effective for creating coproductive relationships as when that connection was paired with the affective interactions that led to personal connections. It was not simply that the technology enabled the staff and mothers to communicate more frequently; it mattered how the technology participated in that communication. When participating as a broker of less-structured messages, the technology created a shared ontological space in which staff and mothers could interact on more equal footing—where the affective interactions drove closer personal ties, created explicit reciprocal interplay between commitments and dependencies, and led directly to new capacities to act for the shelter as a whole. The attachments the staff devel-oped generated new capacities to intervene and work with the residents, and the residents gained new access to information and to contextualizing support for that information that made it relevant and actionable. These new capacities were only possible through the added participation of the technology in the issues of social service provision.

While the technology participated with the staff and mothers—acting alongside with and not simply augmenting their actions—it introduced an experience of personalized information and a heightened perception of personalized care through an environment of shared attachments to com-mon issues, gelling the staff and the residents into a public. These affective ties were created by the Message Center, which acted much in the way other forms of digitally enabled information sharing have acted by creat-ing personalized connections to common causes. Bennett and Segerberg have termed this work "connective action," in which "taking public action or contributing to a common good becomes an act of personal expression and recognition or self-validation achieved by sharing ideas and actions in trusted relationships" (Bennett and Segerberg 2012, 752). By contextualizing instrumental information sharing with the development of personal relationships, the staff and residents and technology were engaged in forms of connective action in which they were not working for a single common identity (as they would under the aegis of collective action), but were instead acting to conjoin distinct but connected individuals and needs.

Sociotechnical Attachments

While issues gave scope to the public at the shelter, attachments created the means to act through a set of shared commitments and dependencies. Given the context of social service provision at the shelter, many of the environmental conditions for developing attachments were already present: the staff were committed to helping the mothers at the shelter recover from a particular crisis and regain independence; the residents depended on the help of the staff to connect to needed services and counseling, health care, and education. Configured in this way, the relationship between staff and residents could easily have been one in which the residents consumed services provided by the shelter; however, as issues were explored and as the experiences with the design process and the technology-based communication practices began to show, a consumption-production relationship between the staff and residents placed artificial limits on the kinds of attachments that might form, imposing a directionality to them that ultimately limited the work accomplished at the shelter.

Developing reciprocal attachments to shared issues and involving both the staff and the residents in creating cooperative strategies for contending with those issues gave substance to the public. The characteristics of these attachments—the complex of commitments and dependencies—exposed facets of the social dynamics that were more nuanced than the rough dyad of care provider and client typically assumed within the shelter setting. Certainly, the staff were part of an authoritative structure, but that authority was under constant negotiation. The design and use of the technology exposed these negotiations by enabling new attachments and new channels of influence with the disclosure and creation of information shared via the Message Center and the Shared Message Board.

The discourse about where and how different forms of information and communication should be deployed at the shelter—for supporting instrumental information needs about programs and services, or doing the affective work of building strong relationships—was enacted through sociotechnical attachments. It was the brokering of commitments to and dependencies on both the social structures between the staff and residents as well as the technical means for negotiating and working within those social structures that made up a machinic assemblage (Deleuze and Guattari 1987). More than just creating a common space in which to act on common issues, the sociotechnical attachments built out a reciprocal set of relations that enabled the staff and the residents to create new capacities to contend with the issues at hand: managing relationships, sharing information, connecting to services, and personal empowerment.

Importantly, affect's ability to describe the interaction of bodies—of people, individual and collective—and the technical relations to and with those bodies is eminently relevant to informing our understanding of attachments and the development of sociotechnical capacities through the constitution of a public. For the staff and mothers at the shelter, membership and participation in the public was bound to coproduction: coproduction of the technology artifact and the practices of use within the shelter, coproduction of information and expertise, and coproduction of social services that were achieved through mutual actions rather than the result of a directional production-consumption relationship.

Such mechanisms of coproduction and the attachments that enable them derive from "[joyous] affect which increases the potential power of bodies, enabling them to 'form new and mutually empowering encounters.' … [This is an important formulation of social encounters because it] emphasises the extent to which the only thing that increases the capacity of bodies is in fact their ability to form productive relations with other bodies" (Gilbert 2013, 147). Here Gilbert is turning to Spinoza's notion of joy or pleasure in which, rather than the result of sating some bodily or emotional need, pleasure is constructed "as the experience of the augmentation of the body's ability to act" (ibid., 183). One consequence of this position is that the affective relations of attachments became the "medium of collective agency and creativity" (ibid., 147), enabling the kinds of coproductive relations between the shelter staff and residents that gave force to the alternative solutions embarked upon by the staff and mothers. Through the constitution of a public, one engaged in designing responses to their issues, the staff and residents developed different affective bonds, bonds that augmented their ability to recognize and respond to the particular issues at the shelter. Conversely, during periods at the shelter in which those affective relations did not materialize, the staff and residents fell back into patterns of producer/consumer relations.

5   Infrastructures and Infrastructuring with Design

Issues articulate the basic conditions around which a public can begin to form; attachments express the affective commitments and dependencies between the actors, artifacts, and institutions contending with a set of shared issues. The third and final piece of the theoretical frame of a public comprises infrastructures and the work of *infrastructuring*. The first two do the work of bounding the design space, first by creating a set of pragmatic points of entry for considering and constructing sociotechnical systems to contend with the issues at hand, and then by establishing a network of social and technical resources that alter a public's ability to act. The final piece comes from the work done to systematically attend to the consequences of a public's shared social conditions by entangling the social and technical capacities for confronting issues with the resources available through, and embodied by, a network of attachments.

Infrastructuring is the integration of social and technical resources through a network of attachments in order to bring future issues into focus and enable a public to become a stable apparatus through which to contend with those issues. Infrastructuring builds on the sociotechnical relations present in community settings and styles the design work to develop capacities to act through those relations. The goal of design in this context is the support of local infrastructures that are reconfigurable by the community so that, as issues and attachments evolve, capacities are in place to manage those future conditions (Ehn 2008). Moreover, it represents an expansive view on what kinds of inventions might be pursued: social, technical, institutional (Björgvinsson, Ehn, and Hillgren 2010; Ehn 2008).

The integration of social and technical and institutional capacities is a key element of why publics are a practical framing for community-based design: by actively seeking to intervene in social conditions, the design process must attend to the relations between actors, artifacts, and institutions and the ways in which those relations constitute and reconstitute themselves. By comparison, where user-centered design and its variants have broadly sought to contextualize design around human activities and their social context, they have too narrowly focused on products for the end user rather than on the capacities that can be built through the design process itself. The focus on products is understandable, given the commercial and professional settings from which user-centered design arises. Having a clear narrative and authority structure within which to operate—that of the workplace, or of the production of products for sale—turns the attention away from the ways in which design can be used to create new abilities to act. The perspective of community-based design shares much with the legacy of participatory design, which has long considered and been driven by understanding the relations of the workplace, the role of practice, and the empowerment of labor to contend with instrumental and organizational issues as a principal factor of design (Björgvinsson, Ehn, and Hillgren 2012; Ehn 1988; Thorsrud 1970). Likewise, the critique that modern design was too focused on the manufacture of products (and the manufacture of desire for products) offered by Papanek (1971), and the alternatives offered by contemporary scholars and designers like Manzini (2015) or Fuad-Luke (2009), all aim to shift the priorities of design practice and sensitize that practice to the many cultural and affective influences that arise through social relations.

Where publics add to these perspectives is through the purposeful shaping of those relations rather than simply taking them as a more or less stable context. It is the development of shared issues and the articulation of attachments that enable a shift from assumptions of stable settings—a move from matters of fact to matters of concern. By focusing on relations in design, we can counter the determinism of user-centered design and contend with the co-construction of issues and responses. With respect to infrastructuring, design is a purposeful act in which new sets of relations are created around issues and attachments—a reconfiguring of the network—aligning different contexts and enabling new abilities to act that simultaneously confront the issues as well as re-create them in recognizable and contestable forms.

For social design, there are two significant implications in the move toward design as infrastructuring. The first is that the work of infrastructuring shifts the focus away from end user products and "is characterized by a continuous process of building relations with diverse actors and by a flexible allotment of time and resources" (Hillgren, Seravalli, and Emilson 2011, 180). Instead of working through design-for-use, developing design programs and processes that support the creation of a product and then step away once the product is delivered, infrastructuring forces design work to be configured around developing a set of capacities that enable design-for-future-use (Ehn 2008; Björgvinsson, Ehn, and Hillgren 2010). The shift in focus from activities that occur prior to use to supporting activities that might occur in the future means that the work of design is no longer about product per se, but instead about creating the conditions in which solutions to future issues can be considered. These conditions arise by working with—and developing new—nodes in the network of attachments. This is what Ehn drove at when he described design-as-infrastructuring as a "sociomaterial public thing, [that] is relational and becomes infrastructure in relation to design-games at project time and … design-games in use" (Ehn 2008, 96). The relationship between the capacities to act that were cultivated during design-before-use (in Ehn's terms, "design-games at project time") and those that persist and are available once the artifact or system or procedure is in use is what I call infrastructuring: it draws directly on the attachments and the affective relations that augment abilities to act; it is directly connected to and motivated by the need to mitigate or amplify the consequences of a common set of issues.

The second implication infrastructuring offers social design is a way to support community practice—the "routines consisting of a number of interconnected and inseparable elements: physical and mental activities of human bodies, the material environment, artifacts and their use, contexts, human capabilities, affinities and motivation" (Kuutti and Bannon 2014, 3545). The turn toward practice is precisely what underpins early Scandinavian participatory design—drawing as it does on Heidegger, Marx, and Wittgenstein—by moving to advance a design agenda built on recognizing the entanglement of knowledge and skill with agency and empowerment (Ehn 1988; Kuutti and Bannon 2014). In laying out the fundamental concept of practice in the opening chapter of *Work-Oriented Design of Computer Artifacts*, Ehn (1988, 60) writes:

In practice we produce the world, both the world of objects and our knowledge about this world. Practice is both action and reflection. But practice is also a social activity. As such it is being produced cooperatively with other beings-in-the-world. To share practice is also to share understanding of the world with others. However, this production of the world and our understanding of it takes place in an already existing world. It is the production of former practice. Hence, as part of practice, knowledge has to be understood socially—as producing or reproducing social processes and structures as well as being the product of them.

Ehn's view of practice again ties directly to the account of practice recently forwarded by Schmidt (2014, 436): practice "encompass[es] not only handling variations and contingencies but [is] also what is done to envision outcome." What these definitions drive at is the recursive way in which relations to issues and attachments constitute certain social structures, that those structures are then reinforced through whatever action is taken in response, and how that response further informs and shapes the appearances and persistence of the issues. In the case of urban homelessness, one such arrangement is of homeless person as recipient of social services. Such an arrangement enacts certain attachments to social service providers and establishes a consumption/production relationship around which revolve the infrastructures of care. Changing that arrangement requires examining the issues that are being addressed, interrogating the attachments, and building new infrastructures to support alternative arrangements. This is precisely the work of infrastructuring: creating new social and material relations between actors, artifacts, and institutions such that a set of shared issues and common attachments produces and reproduces a desired outcome.

By turning to infrastructuring and drawing on *practice*, a term that has been used loosely across different disciplines of computing research, I am casting a light on how design can be deployed in community settings to develop a "practice of community"[1] that attends to the reflective actions of "normatively regulated contingent activity" (Schmidt 2014, 436). In supporting a practice of community, we cannot treat social design as an exogenous influence in which the outcome of the designed artifact has some particular and planned impact on the community. Instead, we must treat design as co-constitutive, situating the specific effects of design within the community's interpretations of those effects. This perspective follows a sociotechnical turn that places technical developments—in this case, community-based design—within a set of social and historical relations (Grint and Worden 1992; Bijker 1995; Latour and Stark 1999). Such a perspective divorces a deterministic view of design (and any consequent technical invention) and instead directs our attention to how any notion of design's impact, whether progressive or regressive, is shifted from matter of fact to matter of concern derived from the social context, the processes of adoption and appropriation and sociotechnical participation, and the cultural and historical realities present for a given setting (Jones and Orlikowski 2009).

Shaping Future Technical Resources

Returning to the scene at the shelter, the design and deployment of the Shared Message Board and Message Center created moments of organizational and work-related reconfiguration for the shelter staff. As the staff adopted the technologies they co-designed, and as that technology participated with the staff in providing social services, changes trickled back into how they conducted their work and how labor was divided between the different caseworkers. The most substantial consequence of this interplay, and one that clearly speaks to the notion of infrastructuring, was the evolution of the Shared Message Board and the corresponding redesign of internal case management documentation.

After the initial design of the technology was complete and the staff and mothers had been living and working with the Shared Message Board and Message Center for a little more than six months, we revisited the design of each. Our goal was to undertake a substantial iteration in order to address the many small adjustments—both social and technical—that had developed over the initial period of habituating to the technology. Where the initial design had sought to address the issues of information and relationship management as they had presented themselves during my earlier fieldwork and co-design sessions, once we deployed the system, the attachments to the issues changed. The staff had come to depend on the technology in new ways and to depend on each other's use of the system to share information and maintain a shared awareness of collective and individual events at the shelter. This dependency began to shift the attention away from the immediate response to present conditions, toward creating a stable platform to contend with future needs and work configurations at the shelter. At the beginning of the project, the staff were narrowly focused on a set of known issues at particular scales of action—sharing and managing information flows, developing new tools for evidence-based care provision, and making regulatory compliance more straightforward and aligned with daily work practices. After working with the system, the staff shifted their attention to building technical and organizational affordances to sustain the information sharing practices they had developed.

The caseworker and program director were specifically interested in creating a consistent information architecture across several activities at the shelter—from the sharing of information via the Shared Message Board and Message Center, to paper-based documentation procedures, to creating kinds of work that could easily be completed by volunteers. The presence of the deployed technology and the development of a rich design-based discourse grounded in understanding and supporting shelter practices helped the staff and mothers at the shelter to think through both the technical and organizational elements of shelter life and to create alternative attachments to the services, procedures, and personnel present at the shelter.

Staff: Infrastructuring the Shelter

The response from the staff that developed through the design process took very specific forms, but each connected to a primary concern around extending some of the benefits they saw in using the technology across the larger organization. The staff wanted to achieve a stable and complementary set of relations among three elements: the sharing of information with the residents, programs within the shelter to connect the shared information with specific courses of action, and a more flexible division of labor to enable the collection and dissemination of more timely information to residents. To achieve these goals, the staff undertook a substantial reorganization of the way they structured and documented their counseling sessions with the residents.

As described in the previous chapter, the deployment of the Shared Message Board and Message Center created a new set of attachments at the shelter, attachments that enabled affective bonds to form, amplifying the capacities to act for both the staff and the mothers. These capacities to act were bound up with the dependencies on and commitments to sharing information and grew out of the capabilities of the technology. As these practices emerged, the staff wanted to reconcile the way they handled the paper-based documentation of individual goals and progress for the mothers at the shelter with the way information was structured and delivered through the Shared Message Board and Message Center. The goal was to align the two sets of practices more explicitly so that it would be easier to coordinate care across shelter staff and so that the staff and mothers could more easily connect the information resources embodied in the technology to the programs and services they were using.

An important component of the shelter staff's decision to reorganize the way they documented services and set goals with the residents grew out of witnessing how the residents responded to content posted to the Shared Message Board and to messages received via the Message Center. By reorganizing the structure of how they tracked services, and integrating that organization into the design of the technology, the staff developed a more explicit dependency on the technology: it amplified their ability to do certain kinds of information work, sharing and broadcasting announcements to the residents and shifting some of the overhead of such information sharing from person-to-person meetings to a channel which encouraged the residents to make use of the technology on their own, thereby supporting the larger goal of empowerment.

After the initial deployment of the Shared Message Board, and in response to discussion and direction from the residents, the staff developed a strategy to align program documentation—and ultimately program structure—with information-sharing services provided by the technology. Alignment took the form of a set of categories—child care, employment, housing, and personal development—around which individual courses of care and counseling could be structured and which could be reinforced via the information services available.

With the new categories for structuring how residents encountered existing programs and how the staff documented progress in those programs, the next move was to mirror that same structure in the Shared Message Board. Doing so enabled a parallel between the information resources available through the technology and the ways activities were structured at the shelter. This desire for more categorized information came not only from the staff but from the residents, who wanted more structure to help make it easier to attend to the kinds of information that was individually relevant (e.g., employment announcements or housing listings or child care information). In the initial design of the Shared Message Board, all information from the staff was placed in a single view, making it tedious to read through a large number of announcements that might not be relevant and difficult to find which announcements were new. By dividing the Shared Message Board into several categories, the staff and the residents reconfigured the technology to better meet the kinds of information sharing practices that had evolved. These categories defined both the common issues and the actions that would arise to address those common issues.

In addition to the new categories in the Shared Message Board, the staff, residents, and I worked together to define a service in which residents could subscribe to particular information categories of interest and receive any new announcements directly on their phone via a text message (in addition to being able to find the information on the Shared Message Board). The motivation here was specifically to address the residents' need to learn what information was new. Moreover, the subscription service attempted to shift the general information sharing into a communication channel that had afforded the development of more personal connections and supported relationship management. These were the positive aspects of building attachments and creating affective ties that enabled new action, and were critical elements that both staff and residents wanted to expand as a set of practices into the future.

These two changes to the technology were in direct response to the shifting ways the staff were experiencing the issues of information and relationship management. Up to this point, the technology had enabled a new set of opportunities for extending the relationships between the staff and the residents, but had largely worked only locally by supporting the attachments that developed within the shelter. As those attachments formed, they created new capacities for action on the issues being faced at the shelter, but begged the question of how similar capacities might be turned toward addressing regional, state, and national issues in which increasingly abstracted regulation relied on similar kinds of information sharing. This in turn led the staff to turn back to the technology and seek ways to redesign and reconfigure the Shared Message Board and Message Center to better support the kinds of future outcomes they were working to achieve: more access to information at both the local and institutional scales, clear connections between the kinds of information provided, and better contextualization so that it might be acted upon. Responding to these needs was not simply about providing the right sets of features in the technology, but about the work of infrastructuring to cultivate practices around the technology so that any temporary benefits would turn into productive long-term tactics for contending with issues across different levels of accountability.

Mothers: Infrastructuring for Support

The mothers had a related set of concerns they wanted to address during the redesign. Their concerns revolved around the structure and organization of information on the Shared Message Board. As pointed out above, a pervasive annoyance of the original Shared Message Board was that finding new information posted to the screen was tedious. Even though each post had a date next to it, messages would scroll across the screen at a modest rate and the mothers would need to wait patiently through all the messages to ensure they saw any new information posted to the board. This factor affected how they engaged with the Shared Message Board: once they had become familiar with most of the information on the board, they either dismissed it as not offering new information or had to spend more time at the screen to look for new posts, often without knowing whether anything new was even available.

Adding categories to the Shared Message Board made it easier to find information about the particular topics of interest and reduced the overall volume of information to sift through. Taken in combination with the subscription service that pushed messages in subscribed categories directly to the mothers' phones, the changes to the technology marked an improvement in the way the system presented information to the residents. These changes were motivated in part by the redesign work from the staff—as discussed above—and in part by the redesign of services by the mothers, who were concerned with creating information services for future residents. Their focus was on creating services that were accessible and not overwhelming and would augment future residents' ability to act on programs while at the shelter.

The redesigned Shared Message Board had an immediate impact on the way the residents made use of posted information. They pointed to the more clearly communicated categories of content as a useful improvement because this reduced the total number of messages in any one category and made it easier to notice new messages in areas of interest. Over time, however, one of the common points of discussion among the mothers was that the information design of the Shared Message Board did not look like something to which they could contribute. The mothers came to view the Shared Message Board as an effective way for the staff to present information to the residents; however, it was not seen as a legitimate space for messages from the residents, as it had been in its previous incarnation. This marked an important and somewhat unfortunate turn in the perception of the technology that accentuated the residents' dependency on the staff rather than cultivating their ability to develop a dependency on the shared knowledge and experience of the other mothers.

The shift in perception corresponds to an alternative set of attachments that emerged through the redesign. In short, the initial design benefited from ambiguity (Gaver, Beaver, and Benford 2003): the lack of extensive categories and the plain appearance of text did not overly suggest specific uses, so the mothers could and did develop a set of attachments to the technology that supported their sense of legitimate ownership of what could be posted to the Shared Message Board. This finding recalls earlier work by Ehn and Kyng in which they used cardboard mock-ups to explore technology design, arguing that the nonfunctional cardboard prototypes created a space for interpretation and adaptation by those they were designed with (Ehn and Kyng 1991).

The redesigned layout, however, with a more structured visual design and more comprehensive content categories, transformed the Shared Message Board's purpose into an unambiguous vehicle for vetted, categorized, and instrumental information. Following these changes, the residents stepped away from the Shared Message Board as a vehicle for own expression and instead developed practices around the information that were tied more directly to the issues and social realities of homelessness that extended beyond their particular relationships at the shelter. It is important to point out that the changes to the Shared Message Board did not come by fiat from the shelter staff, but from the process of the staff and mothers working together and negotiating which elements of the technology supported—and could be refined to continue to support—the needs of residents. It was thinking through future use—use that was not contingent on having cultivated a specific kind of attachment through the design discourse—that allowed the residents and staff to conclude that the most useful configuration was one that focused attention on and contextualized information in order to help future residents more easily take action given the particulars of their situation.

As I worked directly with residents to design the services they had come to use, together we were able to establish and maintain a design discourse centered on the co-option and use of the technology, and to create a dialogue about the broader context in which that technology was situated.

Through that discourse, residents were able to imagine future users—from their own experience and how they would hope to ease that experience for others—and reconfigure the way the Shared Message Board and Message Center fit within the shelter to support those imagined future users. The redesign work comprised an explicit set of infrastructuring actions based on how the future users might respond to the technology, given the understanding of issues and attachments by the present users.

The infrastructuring work done to address future use came at the cost of attachments that had developed during the earlier design process and illustrates, first, that design is a set of tradeoffs in which not every desirable outcome is possible, and second that attachments both amplify and diminish capacities for action. By moving toward sharing information tied to instrumental goals at the shelter, the residents were integrating the technology, creating particular kinds of attachments to available resources, and cultivating practices that would benefit those who would come later. However, the choices they made foreclosed on some of the subjective and social connections that had developed through the course of using the Shared Message Board and Message Center as conduits for building interpersonal relationships.

Once the development team and I deployed the redesigned technology and new and unfamiliar residents came to use it, the technology itself became less contested as a site of active design. Residents accepted it as a given part of shelter life, another tool the staff used to provide information and direct their attention and activities to the programs at the shelter (Le Dantec et al. 2011). Over the subsequent weeks—and now years, as the Shared Message Board and Message Center remain in use—the degree to which the technology became an infrastructure for the residents developed through the choices and forward-thinking strategies of residents at the shelter. Once the technology design (and redesign) stabilized, the work of infrastructuring shifted to the social practices in which the responses to the technology—and the responses to the content conveyed by the technology—were modified and refined across different generations of shelter residents. As much as the technology itself, these social practices of seeking information from the Shared Message Board, using the Message Center to communicate with shelter staff, and sustaining such practices across generations of residents were as much a part of the infrastructure as was the technology, and their perpetual evolution was and remains as much the work of infrastructuring as the explicit technology design ever was.

Volunteers: Infrastructuring New Resources

During the redesign period, a third group emerged as an important but hitherto overlooked constituency of the shelter-bound public: the volunteers who donated time and labor to the shelter. Up to this point, the volunteers had played an invisible role with respect to the design and use of the Shared Message Board and Message Center. During any given week, rotating groups of regular and semiregular volunteers took on different roles at the shelter—meal service, education and tutoring, and some casework support. Volunteers providing casework support came to the shelter as part of their social service training at a local college, and it was this group the staff looked to most explicitly during the redesign.

Since the initial design direction was centered on staff-resident communication, there was a collective assumption at the beginning of the design process that this communication would need to be privileged and would not heavily involve volunteers. However, once the redesign began, and particularly after weeks of working to keep up-to-date information present on the Shared Message Board, the staff realized that it was important to enable volunteers to contribute. They recognized that the issue of providing up-to-date information had spilled over the borders of professional practice and could be extended to other constituents of the shelter. The challenge was structuring those contributions so that casual labor would support the ongoing activities at the shelter and the programs the staff were running with the residents.

Prior to the redesign, the volunteers who directly supported casework tasks spent some time helping with documentation activities. As the redesign of the documentation practices proceeded and the Shared Message Board and Message Center each aligned around a common set of categories, it created attachments with the volunteers by way of the technology without bringing them into direct, and potentially confidential, contact with the conversations and disclosures between the caseworker, program director, and mothers. Instead, the staff and mothers would attend to the categories of information that supported particular programs or goals while at the shelter by subscribing to appropriate categories of information, and the volunteers could work on data entry: adding announcements, links to resources, and new information to the appropriate categories. The information added to the Shared Message Board would then propagate via the subscription service.

What the new categories provided was a way for volunteers to participate in a personalized and contextualized information infrastructure for residents via the announcements and notifications that provided aid in individual circumstances. Volunteers could come and go from the routine of the shelter and still donate work that was valuable to the staff and residents without requiring a great deal of training and without needing privileged access to the residents.[2] The categories and the technology created an effective firewall between the permanent staff and the volunteers—one in which volunteers did not have to be a part of confidential exchanges with the residents, but could still participate in the information sharing that drove much of the work to connect residents to needed services.

By redesigning the entire organizational strategy, the staff, mothers, and volunteers worked to create an infrastructure that more easily supported the evolving constituency of the public, one that would enable new kinds of participation in the production of knowledge at the shelter and thereby increase the capacity of the staff and residents to act. This effort configured the technology and attendant categories of care around relating different contexts (Star and Ruhleder 1996): the context of the shelter staff and how they structured and provided care, the context of the residents as they sought information and connection to relevant services, and the context of the volunteers and how they contributed to the work done at the shelter. The changes that cascaded through the shelter as a result of the infrastructuring work exerted a normalizing effect on each of these different contexts, bringing them into alignment with each other. The normalization enabled the operation of the programs in the shelter and the sharing of information on which those programs relied to be made more robust despite the periodic turnover of both residents and volunteers.

Several pieces came together through the infrastructuring work so that future residents, future staff, and future volunteers could actively participate in the public at the shelter. Binder et al. (2011) have documented several different infrastructuring strategies—component, pattern, ontological, and ecological—which provide different entry points for configuring the design process and the resulting infrastructures that might emerge. As the terms each suggest, these strategies range from a focus on the mutability of designed systems, to the general rules and categories through and over which design might operate, to a systemic or network view of how designed artifacts participate in the world. The relevant strategies deployed at the shelter fall under the ontological and ecological banner, in which the categorization and rationalization of services and information lead to a robust boundary object linking staff, residents, and volunteers in the work of social services. Beyond that, the ecological strategy of infrastructuring can be seen in the way the Shared Message Board, Message Center, and subscription service provided affordances to one another and to the individuals at the shelter so that as issues arose, different components could be enrolled to prompt or support action. This configurability meant that practices could evolve at the shelter, enabling a future public to design responses to issues that may arise.

Shaping Future Social Resources

As the technology evolved through infrastructuring, so too did the social conditions and expectations in which the technology was situated. As discussed above, one of the explicit effects of infrastructuring social resources was the change to how the staff organized the documentation of services. The organizing of services into categories set into motion a shaping of the social setting at the shelter in which staff and residents and volunteers could engage together in a set of practices to address the issues facing the families there. This shaping of practice is, I would argue, the material outcome of infrastructuring. Practices build the relations between issues and artifacts and actors and institutions; they are the connective tissue that makes a public durable, clearly articulated through a network of attachments; they effect outcomes through the social and technical capacities that modify those attachments as the context of social issues or the members' individual particulars evolve.

The infrastructuring work described in the previous section demanded a reflective perspective on the part of the staff and mothers at the shelter. They needed to take their experiences at the shelter, their experiences with the systems of social service, and their experiences with the technology and look toward future configurations of these elements. Such a perspective connects to recent developments in practice theory that point to ways in which social and technical arrangements—infrastructures—"both make practices durable and connect practices with each other across space and time" (Nicolini 2013, 4). The practices at hand were those that enabled and supported shelter programs and the residents' navigation of those programs while emerging from very personal and individual crises. By understanding the outcome of infrastructuring work as a set of practices, we can bind together both affective and technical relations (Nicolini 2013, 85), integrating attachments and infrastructuring to create a stable but adaptive scaffolding on which a public can develop resources to act on shared future issues. The integration of issues and attachments through the work of infrastructuring creates a practice-based identity for the public. Individual experiences imbricated through membership become a durable substrate that supports new members in the public (by choice or circumstance), but leaves the public able to evolve over time. As sociotechnical participation in the public evolves, it reconfigures and redesigns the resources it needs to confront shared issues.

The accretion of experience and mutable identity within a practice is something Nicolini (ibid., 85–86) points to as an important characteristic of practices and how they develop and change:

[Learning] a practice involves, by definition, a conflict between continuity and displacement: "newcomers are caught in a dilemma. On the one hand they need to engage in the existing practice, which was developed over time. … On the other hand they have a stake in its development as they begin to establish their own identity in its future." … The ensuing conflict is what puts "practice in motion" (ibid.) and what makes change one of its fundamental properties.

By establishing a set of common artifacts that support and constitute a set of practices, the engagement with those practices is re-created and incrementally changed over time and over space, slowly adjusting to new conditions and constituencies. One critique of this model of practice is its lack of capacity for radical invention—a critique aimed particularly at Lave and Wenger's (1991) notion of communities of practice. The critique runs that the incremental change effected by the ever-developing identity of a public does not explain radical invention within the public (Nicolini 2013, 92). How, if the practice is stable, supported, and constrained by social, institutional, and technical capacities, can such invention (rather than measured, incremental change) arise?

The response to this critique is that from the perspective of design and a designing public, it is the work of infrastructuring itself that leads to invention through the response to issues, through an evolving set of attachments, and through the explicit work done to enable change and adaptation aimed at future use rather than only focusing on present situations. Design for present issues indeed leads to the kind of incrementalism that refines present conditions and experiences, but this is distinct from the invention that can create a radical new formulation of a solution, or more likely a radical reformulation of the problem. This tracks the recent critique of practices like user-centered design, in which the notion of invention (or innovation) is overvalued and misapplied given the strengths (and limitations) of what user-centered design as a method actually enables: incremental improvements and inoculation against spectacular failure (Norman 2010; Norman and Verganti 2014).

In the case of the shelter staff and residents, incremental improvements to the way they contended with issues like relationship and information management came through the initial design, which took as its challenge dealing with the present use of technology and information-sharing practices at the shelter. The changes offered by the Shared Message Board and Message Center were substantial but incremental improvements arrived at in response to present issues. It was only after developing new practices with the technology and the shifting of work and responsibility that the staff and residents began to cultivate a sense of how to further redesign the technology and the social processes so that future issues and future uses (and future users) would be readily supported. What is important to point out here is that the work of infrastructuring, and the scaffolding work necessary to arrive at a point at which infrastructuring through design was possible, required developing a set of capabilities and capacities to contend with future use. These capabilities and capacities were, in many ways, orthogonal to the sociotechnical practices within the public that the infrastructuring work was meant to support. Many of the design choices that were attributed to infrastructuring work, and to stabilizing the public by considering how the technology and organization at the shelter would support future generations of shelter residents, resulted in undermining some of the practices that developed during the design process—the reflective use of the technology to build strong relationships between the staff and residents being the most visible casualty. What resulted instead was a set of features aimed at supporting staff, residents, and volunteers that did not assume the same kind of close collaboration that developed through the course of the design project, but that also did not foreclose on such reciprocal relations developing in the future.

Evolving Practices among the Staff

In the first phase of the deployment, the staff developed practices around the Message Center focused on communicating directly with the residents. The majority of the messages sent by the staff were individual communications concerning the particular circumstances, programs, and goals of each resident. With some of the social interactions moved into a technology-mediated space, the residents had more dynamic access to information from the staff—e.g., they could receive information while away from the shelter—but more importantly, the residents also gained an ability to manage their responses to that information. By having information sent to mobile phones, the residents could now make a decision about what information to act on and what to ignore without direct confrontation with their caseworker. This change impacted how the staff approached conflict with the residents and how they negotiated confrontation among themselves: whereas they embraced using the technology to send messages about resources, services, and coordination for the autonomy it provided, that same level of autonomy was met with skepticism when applied to the enforcement of shelter rules.

After redesigning the technology and the overarching organization of how staff-resident interactions were structured and documented, the staff altered their practices with the technology. In particular, the categories of information and the subscription service changed much of how the staff used the technology: the apparent ease of subscribing residents to categories of interest became a shortcut to ensure that information was being sent to those who needed it. At the same time, there was a shift from personal messaging to simple broadcasts of information. The infrastructuring work that was done to stabilize and normalize the sharing of information across several media— the Shared Message Board, the personalized messages via the Message Center, and the overall structure of services at the shelter—shifted the legibility of the system from one built for personal communication between staff and residents to one built around information broadcasts. Whereas the initial design led to conversations between staff and residents, micro-coordination about services and appointments, and a more expressive rapport between staff and residents, the redesigned technology did not lead to phatic interactions between staff and residents, and as a result did not support building rapport nor the micro-coordination that marked the first phase of the deployment. In short, as the attachments to the technology and mediated by the technology evolved, certain abilities to act were diminished.

In place of the design-centered attachments, however, was a set of technology services and social practices that enabled the staff to send contextualized information to residents that was not contingent on working with future residents in a design-led program, but instead was meant to address the ongoing reality of crisis and complex individual circumstance. The changes we made together to the technology provided stability for the programs at the shelter so that they would better fit the roles and responsibilities divided among staff, residents, and volunteers. By posting information to the Shared Message Board or sending it directly to residents via the subscription service, the shelter staff were no longer case-by-case gatekeepers to information in the way they had been. Nor were they cultivating the closely held relationships via the technology as they had been.[3] For example, staff or volunteers could make information about housing programs or job opportunities available to everyone in the shelter via several redundant communication channels. In some cases, the Shared Message Board meant anyone had access to the information; in others, the staff sent information individually via the Message Center. The result was that the staff could more easily manage their interactions with residents and help them navigate the programs and services available. The technical infrastructure of the Shared Message Board, Message Center, and subscription service met with and reinforced the infrastructures of practice that developed around sharing information; they complemented each other and enabled the staff and mothers to extend their limited capabilities both by mediating information sharing and enabling additional work to be completed by volunteers. Importantly, the outcome of the infrastructuring work was a set of practices that were built around an understanding and expectation of the contingent work that would occur going into the future: staff members might change or take on new roles, volunteers were certain to change, and residents would continually cycle through the shelter's emergency services. The approach the staff had taken was to work through these contingencies with as much foresight as they could in order to design stable technical capacities and professional practices that would enable a durable public to contend with issues of homelessness.

Developing Identity and Practice among the Residents

Creating a complement of durable practices among the residents was a different kind of challenge. Whereas the staff were building on a history of providing services and working together, the rapid turnover of families at the shelter meant that new groups of residents were constantly arriving and had only a short time to acclimate. Moreover, the urgency of their stay exacerbated the need to quickly become initiated into the practices at the shelter and to effectively become a member of the sustained public, all while contending with the issues of homelessness and establishing a network of attachments needed to begin to overcome the immediate crisis.

During the course of learning to use the initial features of the technology, the residents developed a set of practices that had a substantial impact on their ability to work with information and build out a set of attachments at the shelter. As detailed in the previous chapters, those capabilities were closely tied to the way the technology blended personal and group communication. On one hand, the Message Center created a channel for the staff and residents to work very closely together on developing the information resources needed and coordinating around those resources. On the other hand, the Shared Message Board became a focal point for announcements and shelter-wide coordination (and issue sharing). Both enabled asynchronous communication to develop at the shelter and gave rise to a set of practices whereby residents became more active in seeking and responding to information as well as developing a more specific sense of what kinds of information would be most valuable.

The sense of what kind of information was most valuable led to explicit design choices by the residents to capture and promote information that matched their priorities. The categories of information that rolled through the entire shelter organization were the result of the residents and the staff co-developing their understanding of how to make the most of limited time and attention to ensure that access to needed information was matched with guidance and direction on how to take action on that information. As pointed out earlier, the work that went into developing the categories and moving the technology design to better support distributing information along those categories was a specific form of infrastructuring that resulted in a boundary object relating multiple elements of shelter life.

Once the redesigned Shared Message Board and the subscription service were put into use, the issues that residents experienced circulated in a different way. The mothers who entered the shelter in the following weeks and months habituated to the technology in a much less personal way; they no longer used the Shared Message Board to organize around common grievances at the shelter, and the phatic interactions that had developed via the Message Center became less frequent, replaced by more instrumental messages via the subscription service. This shift from practices that cultivated personal connections between the staff and the residents to practices that were predicated on receiving information stemmed from the altered way the technology participated in circulating issues—it had transformed from an open platform, legible as a kind of shared bulletin board for everyone at the shelter, into a more directed information interface closely tied to the operations and organization of programs at the shelter.

This shift in how the technology was legible and how it participated in shelter life encapsulates an important point concerning publics and the work of infrastructuring through design. The kinds of practices that enabled the residents to engage in infrastructuring—to consider and work toward addressing future use and future issues—required a social connection and robust engagement with the public at the shelter. These practices did not simply arise out of seeking services, but from participating with the staff, other residents, volunteers, and mediating technologies to coproduce how those services were or might be enacted. The development of attachments to the social and technical resources at the shelter helped the residents act in new ways when redesigning the technology. The perspective they gained and the capacities for design came in large part from the information and organizational practices that had emerged. These practices included a sense of ownership and legitimate use of the technology at the shelter—they were working to design it and make it better, not for themselves but for future residents. The kinds of capacities they developed to be able to consider those future residents were different from the capacities they—and the staff—sought to support with the new configurations of social and technical infrastructure they were creating.

The residents benefited from the way the technology initially supported personal connections with the staff (Le Dantec et al. 2010; Le Dantec et al.

2011)—in fact, satisfying a documented preference for personal communication, building relationships, and developing a confidant to help manage care (Alexander et al. 2005; Beegle 2003; Hersberger 2005). However, when they were considering future use of the infrastructure, the residents' work on the redesign focused on the information practices that were more explicitly connected to institutional interactions. The mothers felt that those practices had the greatest impact as they worked to reestablish stability in their lives. The shift in part traces the utilitarian view they had developed of the technology. Once they were habituated, the Shared Message Board and Message Center became useful components of their time at the shelter, and while the technology supported building relationships between the staff and the mothers, it did so invisibly. Meanwhile, the instrumental and programmatic information shared via the technology was more overtly connected to the outcomes that both the staff and the mothers wanted to achieve. The result, as the mothers considered future residents of the shelter, was that they focused on the information services and left the relationship building as a by-product of personal chemistry between the staff and future mothers.

The redesigned Shared Message Board and subscription service both met the goal of emphasizing instrumental information sharing, and a set of practices took hold among the staff and residents and volunteers that first and foremost promoted sharing information about programs and services in a way that was contextualized for each of the mothers individually. However, as new practices developed and the technology evolved, one of the things that changed was the perception of ownership; when the residents felt ownership over the technology, as they did during the initial system design, there was a clear path to being able to change the technology.

Within participatory design, there has long been a concern with different aspects of ownership in the design and development of artifacts and systems (Balka 2006). These concerns are particularly relevant in community-focused endeavors and social design in which a sense of ownership of the final product has been found to be critical for project sustainability (Carroll and Rosson 2007)—a point that is only supported through the different modes of use observed across the changes to the system deployed at the shelter. There is a distinction to be made, however, in that what is perceived to matter for infrastructuring is not the ownership of the material product so much as having a stake in shaping future attachments by way of a relationship to the material product. Viewed this way, the residents at the shelter had stronger notions of ownership of the technology during the first phase of the deployment in that we were then collectively continually reimagining its use. Following the redesign, as the notion of ownership and perceptions of legitimate right to co-opt the system faded, the way the residents viewed the technology as a resource changed because the technology was no longer the primary focus; rather, the information it mediated was.

Infrastructuring through Design and Practice

As noted at the beginning of the chapter, one case for infrastructuring in design entails a shift of focus from design-for-user to design-for-future-user; or perhaps, more simply, from product design to metadesign (Ehn 2008; Binder et al. 2011). The argument for metadesign that Binder et al. and Ehn have put forward develops out of the notion of *things*, which covers the small-*p* political aspects of objects in the world: the relations between actors, artifacts, and institutions and the ways in which those relations respond and give rise to matters of concern (Binder et al. 2011; Latour and Weibel 2005; Latour 2008). Building on these relations, "the outcome of the design process is a *thing* that modifies the space where people live: besides and beyond its functions (living for houses, hosting artworks for museums, sitting for chairs, etc.), the designed *thing* aims to change the experience of its users; it is rich in aesthetical and cultural values, opening new ways of thinking and behaving" (Binder et al. 2011, 51). This exposition of *things* and their relation to design is narrowly constructed around architectural design and ignores the ways that many kinds of designed artifacts and systems, including those that result from engineering design practices and those embodied in interactive technologies, reconfigure the ways we know about the world. However, the assertion stands that for *things*, the point of departure is the set of relations between actors and artifacts and institutions and the ways in which those relations exist in a constant state of flux: the issues at hand and the attachments to those issues arise from and give rise to the *things* themselves.

Where this matters for infrastructuring is the focus for the outcome. As Binder et al. (2011, 157–158) put it in relation to participatory design:

Participatory design is considered as an approach that tries to involve users in design, and, in this way to encounter in the design process what Johan Redström (2008) has characterized as "use before use." … [Infrastructuring] has to do with how to defer some aspects of design until after the design project is completed, and opens up the approach of use as design, of "design after design."[4]

Another way to put this distinction is that participatory design, like other forms of product design, has focused on the creation of products that, once developed, are handed off to the consumer and no more is done with them. Infrastructuring, by contrast, instead of focusing on the product, sets about "identifying, designing, and supporting social, technical, and spatial infrastructures that are configurable and potentially supporting of future design games … [in which] the purpose is to produce a potential *thing* that will be open for controversies from which new objects of design can emerge" (ibid., 172). It is precisely because the final users are unknown that infrastructuring must look toward building different kinds of capacities—social, technical, organizational—so that as future users are pulled into the context, the design artifacts and systems are stable but remain mutable and open to reconfiguration and redesign as the issues and attachments evolve.

Even as Binder et al. differentiate infrastructuring from participatory design, they note that participatory design in its earliest forms was "not primarily designers engaging in use, but people (as collectives) engaging designers in their practice" (Binder et al. 2011, 162). So whether we give it a special name or simply adjust the focus on existing design methods, the work of infrastructuring means attending to a distant horizon in which design is not a privileged activity undertaken by specially trained individuals, but instead an ongoing act of appropriation and capacity building through social and technical relations (Fischer and Scharff 2000). Infrastructuring is a way of setting the stage for this perpetual design and redesign through use. It is a result of particular arrangements within a public to develop durable resources that marshal attachments to contend with issues, and it enshrines an outcome by establishing a context in which that public is made durable by way of a stable set of relations between actors, artifacts, and institutions.

The practices that developed around the Shared Message Board and the responses to information shared via the technology were part of developing a shared identity and established a link to the public at the shelter. This identity was contested at different points in the design process. It was contested as work responsibilities and influence shifted among the staff; it was contested as the value of the technology moved from supporting interpersonal connection to driving information-seeking activities among the residents; and it was contested as volunteers were enlisted to use the technology and participate in the information-sharing and service-documenting practices at the shelter. The infrastructure of the technology and social interactions around the technology supported and constituted these practices, and as generations of residents cycled through the shelter, the practices and uses of the technology evolved. At first the technology was a curiosity, something a few residents used when communicating with the staff. After a time, it became central to the communication between staff and residents and led to a significant reconfiguration of work and authority at the shelter. Finally, the technology focused less on the discursive relationship between staff and residents and evolved through specific work done by staff and residents and volunteers, to exist as a resource for the particular relationships among them.

Over the course of the project, the aim was to co-develop a set of technologies that would have lasting value to staff and residents of the shelter. It is important to point out two conflicting goals that arose from this desire. In the first instance, there was a need to develop over time a set of capabilities that would enable a design-based project to proceed. In effect, this meant creating a set of attachments within the shelter that focused on designing, using, and then reflecting on the deployed technology and questioning how well it would work for future users who would not have come through these focused design workshops and discussions. These particular attachments to design—to systematically approaching a complex and contingent setting (Cross 2001; Nelson and Stolterman 2002)—created new capacities for action, which were reflected in the design decisions taken during the redesign. From a second point of view, however, the infrastructuring work aimed not to make designers out of staff and residents, but to sustain the programmatic goals at the shelter as a hub for support and social services for future families confronting homelessness.

The design of the communication platform and the development of particular information-sharing practices around that system are not the final culmination of the public; they are simply one step in a recursive process in which the determination of issues, articulation of attachments, and work of infrastructuring resources and relations perpetually constitutes that public, making it durable but not fixed, a sensible context for design but not an a priori locus for a solution.

6   Designing Publics

Issues, attachments, and the work of infrastructuring are the entangled components that constitute a public. Together, they provide both purchase for considered analysis of sociotechnical configurations in community settings and footholds for structuring and advancing the generative action of social design. The account I have provided in this book examines how a public took shape over time and across multiple sites through the process of designing technical artifacts, organizational procedures, and social practices. The constitution of the public at the shelter emerged from the exploration of issues and the relations of attachments, which then became a boundary for infrastructuring work grounded in an evolving set of sociotechnical capacities and practices. While my account of the process is linear and progresses through a timeline of the design intervention, the reality is that working through these elements was a recursive process in which I and the staff, volunteers, and residents would work through issues and attachments and infrastructures to arrive at a new understanding of the shared conditions and a new set of strategies for contending with those conditions. It is in this way that the public was both designed and engaged in design.

The public at the shelter was designed to the degree that the configuration of services, procedures, and supporting technologies established and stabilized the sets of relations between the staff and the mothers at the shelter: it was the result of considered design moves taken collectively to incorporate the needed resources and capacities to overcome homelessness. The focus on sharing information and documenting interactions in ways that consistently reinforced the sources and kinds of information present placed a set of constraints on how the public formed and how it took action to contend with the social conditions of homelessness. Within these constraints, however, the public continued—and continues—to design resources and processes capable of bringing together care providers, volunteers, and homeless mothers in the common enterprise of alleviating the symptoms of poverty and homelessness while trying to establish sustainable responses to the individual and systemic causes of same. While the design project concluded in 2011, I have continued to work with the shelter, occasionally needing to "kick" the technology after it falls over or otherwise assisting with system maintenance and working out small technical changes to adapt to shifting priorities at the shelter. Throughout that time, the staff, mothers, and volunteers have continued adapting their processes and redesigning the technology to accommodate the issues—organizational, social, and technical—confronted daily.

This dual nature of a public—that it grows out of a considered program of design, a sociotechnical assemblage constituted in response to specific conditions, and that it is actively engaged in designing itself, selecting and building capacities to contend with those social issues—illustrates the analytic and generative stance publics entail as theoretical perspective for design. The analytic view allows design—in particular design within the many areas that constitute human-computer interaction—to constructively articulate sites for intervention and action. Whether through issue formation, the articulation of attachments, or the work of infrastructuring, each provides analytic purchase on community-sited design. It is not necessary that every design should holistically tackle each element; rather, they provide entrance and exit ramps that afford a clear and systematic way of relating present and future use as a design intervention develops. In this way a public might be designed, through selecting issues and informants from affected communities, making choices about how and where to intervene and what kinds of participation to seek out and elevate.

The diversity of participation is an important point to highlight here, in that a public assumes a range of participation and does not prescribe specific roles or responsibilities. Paralleling the point McCarthy and Wright make in their recent book on modes of participation in design, participating in a public through design is not about making every member a designer, but about building a venue for genuine exchange and understanding (McCarthy and Wright 2015, 20). Publics accommodate multiple subjectivities as their members act in response to shared social issues. The experiences of individuals, artifacts, or institutions each have a role to play, and it is the agglomeration of these roles that gives a public its characteristics and its capabilities to act in the world.

The focus on building capacities to act through design, and specifically the move toward collective action, transitions publics from an analytic position to a generative position in which the communities constitute themselves as a public in response to shared issues. Like so many computing artifacts, publics are never complete, nor are they born whole through a priori analysis and boundary making; rather, they evolve over time, through participation and action, clearing their own path by way of the sociotechnical materials and relations they create. The design responses to shared issues that a public cultivates fall under the longer-standing observation that design is only ever "completed" through use, appropriation, and ownership (Carroll 2004; Balka 2006; Merkel et al. 2004). By moving through issues, attachments, and infrastructuring, a public develops collective capacities that are grounded in an ownership of both the issues to which they are responding and the means of response to those issues. Further, publics do so by including multiple subjectivities: they are not composed exclusively of user, consumer, or producer; instead, publics work under the assumption that use, consumption, and production are all co-created and in circulation with each other.

Together, the analytic and generative positions fuse and support a productive framing of what I earlier termed *social design*. I want to return to this term now and more fully sketch out the boundaries of social design and the operational linkages between it and publics. Threaded through the book has been an inclination toward reimagining forms of collective action and the role that design might play as means toward such action. The three conceptual pillars of publics each rely on different facets of collective action: first, *issues* create a ground upon which a collective might form; second, *attachments* form the connective tissue of a collective; and finally, *infrastructuring* describes the collective labor that draws upon attachments in order to address issues. Design's role in each facet can be the work itself—in which a public engages in design to address a particular issue—or it can be the initial push that articulates issues and attachments.

Design in Public

As a site for social design, publics tie into a particular kind of collective action, one that is rooted in a common community, bounded by shared experience and shared place. My attention to place is rooted in an expression of urban computing in which the city, in its many forms, hosts the intersection of design and computing and communities. This focus follows my interests and should not be taken as a constraint on the notion of publics, nor on the sites in which consideration of issues and attachments and infrastructuring might be fruitful. The digital diaspora that interacts through social media, or the communities that discover each other through online fora, each articulate common social issues, articulate attachments across a range of social and technical resources, and engage in infrastructuring work to develop and sustain responses to those issues. What follow from these varied sites, whether tracing global political upheaval (e.g., Shirky 2011) or the appropriation of computing to local cultural and economic needs (e.g., Burrell 2012), are practices that reflect place, and publics whose identities are irrevocably tied to that place.

The boundary that I do want to trace encircles a kind of participation that publics entail and that motivates a desire for a social design agenda. McCarthy and Wright (2015) have recently given an excellent account of how participation unfolds in computing's many locales, working through the ways relationships and community membership set the groundwork for participating in different kinds of publics. Where I have narrowly considered the notion of publics from the legacy of Dewey, McCarthy and Wright look more toward Warner (2002), and interweave Hauser (1999) to establish a set of criteria for vernacular participation in computational, or computationally mediated, normative publics. Drawing on their earlier foundational work on experience design (McCarthy and Wright 2004), the criteria they advance for participation enable a broadly inclusive framework rooted in the multifarious experiences of participation enabled, mediated, and amplified by computing. However, where their focus is on the aesthetics of experiences as people perform participation in normative publics, I would draw attention to the ways publics shape participation, not through interactions performed with computing, but through design and the intentional development of capacities and practices to transform a set of social conditions.

This focus on the social conditions is a key boundary. What McCarthy and Wright (2015) provide is a way to carefully think about participation and human-computer interaction—whether by integrating participation with mobile phones at a Rihanna concert, or through a digitally mediated national project to weigh plans for rebuilding after 9/11. While participation is a necessary precursor, one that needs care in definition and understanding, what I am chiefly interested in here is how design operates through participation in issues, attachments, and infrastructuring, constituting a public and confronting a set of social conditions. The previous chapters have endeavored to explicate and ground where and how participation and design and computing intersect to create the conditions for empowerment and emancipation. I chose to ground that account in a single, long-term case study in order to holistically explore the recursive linkages between issues, attachments, and infrastructuring.

Embracing publics as an analytic and generative frame need not be a totalizing position, but can be deployed piecemeal and over time. As community-sited projects develop, and as an agenda of social design begins to take shape, there are a number of different conditions in which publics can help frame and provide trajectory and working model. At the Digital Media program at Georgia Tech, there are a suite of ongoing projects run by myself and my students that connect into interests of civic media and grow out of explicit grounding in combinations of issues, attachments, and infrastructuring.

Cycle Atlanta

The first is Cycle Atlanta, a project in collaboration with the City of Atlanta and the Atlanta Regional Commission in which I, my colleague Kari Watkins at Georgia Tech, and our students released a smartphone app that lets citizens record their bike rides. By analyzing route data, ride purpose, and rider demographic data, city transportation planners can build a robust empirical foundation for prioritizing and targeting new cycling facilities in the city.

The deployment of the smartphone app channeled two streams of contemporary optimism on how digital technologies can transform urban life and civic participation. The first drew on the vision of smart cities projects in which sensor networks (e.g., Vakali, Anthopoulos, and Krco 2014; Olivares, Royo, and Ortiz 2013), instrumented infrastructure (e.g., Erickson et al. 2012; Perera et al. 2014), and participatory sensing (e.g., Roitman et al. 2012; Doran, Gokhale, and Dagnino 2013) are interwoven to improve operational efficiency in city services and planning (Perera et al. 2014). The second drew on moves in digital democracy to increase participation by augmenting established face-to-face interactions with asynchronous, situated, and locative modes of interacting with public officials and policymakers (Hacker and van Dijk 2001).

Since its launch in 2012, the app has been picked up by parts of the cycling community and helped a small but influential public gain traction when advocating for additional cycling facilities in the city. Like the case of the homeless shelter, in which the design program began by examining the organizing *issues* around which a public might converge, the Cycle Atlanta project focused on articulating and leveraging *issues* within a cycling public primarily engaged in cycling as transportation—commuting to and from work—rather than as an activity motivated by fitness or sport. The data that were produced by cyclists make this public visible both to itself and to the city as governing entity through artifacts like interactive maps and public reports (Alta Planning + Design 2013). In turn, this visibility enables the public to articulate specific issues: issues with the specific physical conditions of cycling in the city via shared experiences on specific roads, intersections, or traffic conditions; issues driven by a community of cyclists that is otherwise more difficult to identify due to its small overall numbers and the diffuse manner in which cyclists navigate Atlanta; and issues with political advocacy for resources and modification of the built environment.

The work of infrastructuring between cyclists and transportation planners is the focus of the project, but it remains nascent as we continue to explore both the computing tools needed to make sense of these new data and the sites where that sense making takes place. One such site focuses on finding ways to circulate the ride data within an ecosystem of visible production, use, and repurposing by members of the cycling community. The naive aim here is to begin by exploring ways the data produced by the community can support the locally developed practices of cycling through the city—sharing knowledge about navigating the city, making patterns of traffic flow legible, enabling collective forms of organizing and acting to arise out of individual and private modes of data collection (via personal devices tucked away in pockets and bags). These efforts build on the issues expressed through participation in the public via use of the app—they provide a pragmatic way of identifying individuals concerned with the conditions of cycling as a form of transportation and a fulcrum around which to leverage those concerns, and, by doing so, bring a collective—a public—into being (if not an ability to create consensus or make collective decisions) (Gilbert 2013, 24).

These efforts also key into the multiple subjectivities circulating within this particular public of cyclists. There are a range of experiences in cycling through the city: not everyone who rides a bike identifies as a cyclist; not everyone who is a cyclist identifies as an advocate. Accommodating and supporting these different experiences through design begins by understanding and enabling attachments to form across a diverse range of individuals and includes not just the ways people identify with cycling but ethnic, socioeconomic, and cultural identities as well.

The second site of interest concerns the professional practices of planners and city officials who have incorporated the ride data into their decision-making and urban design processes (Le Dantec et al. 2015). Here, the issues thread through the ways in which planners can draw on community knowledge to make informed decisions, and how they can use data produced by cyclists to support particular transportation design decisions. In this way, the app and the data participate with segments of the cycling community and the city's planning apparatus, joined through a set of attachments that augment the cycling community's ability to act as well as supporting the professionals in their day-to-day decision making. The sociotechnical relations between cyclists, their data, and planners give way to new issues that arise with respect to equity of participation, veracity of representation, and accountability and transparency of civic institutions.

While the articulation of attachments plays a role here, it enters as a secondary concern. Instead, the project revolves around the formation of issues and the work of infrastructuring through design interventions to create and sustain the app. The distributed labor of collecting and visualizing the data so as to build connections across individuals committed separately to cycling as a transportation practice, to forms of grassroots civic advocacy, and to particular visions of urban renewal through the built environment are the outcomes of this focus. What these efforts begin to lay bare in an explicit and observable way is that "service outcomes are the product of collaborative and creative relationships between professionals and members of the wider public" (Gilbert 2013, 169): that the relationship between city and citizen should not be narrowly construed as one of service and consumer, but one in which the two reproduce each other through their attachments and the ways those attachments augment abilities to act on the issue of cycling facilities. Similarly, the relationship between care provider and mother at the shelter was better viewed through the attachments in which social services and their outcomes were coproduced through the attachments each had within the public of concern.

Community Historians

A second multiyear collaboration with a group called the Historic Westside Cultural Arts Council began with expressing attachments across individuals and institutions in a distressed Atlanta community. The work we have been doing falls under the umbrella name Community Historians; in contrast to the Cycle Atlanta project, in which issues and infrastructuring were the main work of the project, the Community Historians project was concerned first and foremost with developing a network of attachments through which new sociotechnical infrastructures might support community action and new opportunities for civic engagement. Like many distressed communities across the United States, the neighborhoods where the project was sited were facing an assortment of deeply rooted and interconnected social issues: high unemployment with a large portion of the community living at or below the poverty line, high vacancy rates in community houses and commercial buildings, very low rates of owner occupancy, inadequate and failing public infrastructure, and the attendant challenges of drug dealing and use, prostitution, and petty and violent crime. These issues were only compounded by the most recent downturn triggered by the bust in the U.S. housing market—many of the properties in the neighborhood were tied up in institutional ownership and speculation that ramped up during the boom. When the housing market collapsed, those same institutional owners were not members of the community and let their properties languish, further exacerbating challenging conditions in the area (Mariano 2014).

The issues around which the community might organize and take action were multiple. However, a component of the challenges was what could be called low social capital, both in terms of how the community organized itself—bonding capital—and how it connected to and activated resources externally—bridging capital (Putnam 1995; Larsen et al. 2012). As a design problem, developing these twin kinds of social capital can be seen through the lens of articulating and cultivating a set of attachments to shared issues within the community and to different social and technical resources in the community's immediate orbit.

The project was aimed at building tools and technical capacities with the community to address these issues. Documenting local conditions, facilitating collective decision making, and enforcing accountability each present concrete opportunities for specific local invention rooted in community expertise. While the work to enable the community to confront these issues looks ahead toward forms of infrastructuring across individuals, artifacts, and institutions, the more pressing project of the design program is to identify and cultivate attachments across a public. These attachments are grounded in making visible a shared identity rooted in the cultural and historical legacy of the neighborhoods—a legacy with bright-line ties to the birth of the civil rights movement in the 1960s. It is through this shared identity that my community partners in the Historic Westside Cultural Arts Council focus on human development as a desperately needed but invariably overlooked complement to the economic and real estate development projects currently sited within their neighborhoods.

By focusing on human development, the Community Historians project circles back to social design and the ways in which attachments and affective connections augment a community's capacity to act. The project involves computing by exploring ways of digitally documenting the social and physical experiences of the area (Fox and Le Dantec 2014); but the aim is not to produce new forms of computing per se, but new relations between the individuals and institutions that participate in the neighborhoods. These relations, understood as a collection of attachments, become a point of departure for the infrastructuring work that relies on local capacities to act and respond to changing conditions. They also help reframe the issues being experienced in the community (Le Dantec and Fox 2015), recognizing that "fixing" crime, poverty, and un- and underemployment are not individual issues that can be solved via individual actions and individual accountabilities, but need a collective force of will that requires attending to human development and a community's commitments to and dependencies on itself. Further, it requires recognizing the complex subjective experience of belonging to this particular community.

The Community Historians project seeks to build new capacities to act by entering through the practice of community—the normative and moral commitments to place and to the people who inhabit that place. The challenge here, as with the homeless mothers, is developing a shared sense of identity that helps link individual experiences with common action; but while the experiences of the mothers were made similar by the acute emergency of arriving at a homeless shelter, the experiences of the individuals and families living in the struggling Atlanta neighborhoods do not share the same acuteness and urgency. Instead, their common ground is built on the bedrock of the local legacy of the civil rights movement, but has been eroded by long-term neglect—in schools, in employment, in every manner of social justice. Developing a shared identity, then, means linking current individual experiences to the recent past, in which strategic community action brought about substantial local and national changes, and in which the commitment to contesting unacceptable social conditions was complemented by dependencies on a network of actors, artifacts, and institutions that enabled activity to turn into action.

Activist Computing

Finally, with Mariam Asad (a PhD student at the time of writing), we have been working with social and housing justice activists for the past three years to understand how situated practices of nonviolent resistance intersect with communication and coordination infrastructures available through computing and social media platforms. The work has concerned itself most directly with the temporal, social, and political constraints of working outside institutionalized channels of action—through street protests, sit-ins, civil disobedience, and forms of noncooperation rather than through lobbying, seeking bureaucratic remedy, or relying on electoral politics for change. Within this setting, she has been carrying out a series of design workshops to explore how these activists frame the short- and long-term issues they face (Asad and Le Dantec 2015; Asad, Fox, and Le Dantec 2014).

Within this venue, a public has emerged around a set of issues related to housing justice and the government and financial conditions that led to a widespread housing crisis following the 2008 U.S. recession. In an arc that shares many of the same contours as the relationship between the shelter staff and the homeless families, the activists and citizens with whom they are working are likewise linked through their attachments to the issues of housing justice. Homeowners invariably have individual and unique circumstances and relations to housing issues that lead them to seek out the aid of the activists; once working with the group, however, the issues coalesce into a set of common hurdles around which the activists have organized services and aid programs.

Working with homeowners to avoid foreclosure expresses a set of attachments—commitments and dependencies—within this public. Again, tracing the contours of the homeless families, the activists are committed to the social and moral obligation of helping distressed individuals and families retain their homes. Meanwhile, homeowners develop a set of emotional and material dependencies on the activists, relying on the organization for help and expertise in navigating financial institutions and legal options for keeping their homes. Further, the homeowners also depend on the network of activists for emotional support if evicted—relying on the affective bonds that arose from shared commitments.

Our involvement with this particular public has been to develop a set of design-based interventions that build on well-expressed social conditions and a clear network of attachments in order to help the activists stabilize their activities and membership across services and forms of nonviolent protest. Grounded in infrastructuring work to more concretely incorporate computing capabilities into the ad hoc activities tackled by the activists, our design program draws on critical making and design to connect the situated and everyday issues confronted in protest actions to ways in which imaginative and speculative technologies might participate (Ratto 2011; Asad, Fox, and Le Dantec 2014). These workshops created a point of departure for thinking about the different ways in which computing was already present in the daily routines of the activists and how they might construct a stable set of tools to support ongoing activities and organizational norms.

By starting our work with the activist community with interventions aimed at infrastructuring, we were then able to use those activities to reflect back on the issues and attachments circulating within the public and seek new ways of broadening enrollment. The result was a nascent understanding that even though the activists had cultivated an identity around public protest and civil disobedience, much of the everyday issues they faced stemmed from the kinds of mundane organizational management issues seen across nonprofit service organizations (Le Dantec and Edwards 2008b; Voida, Harmon, and Al-Ani 2012; Voida, Harmon, and Al-Ani 2011). These mundane tasks required the enlisting and sustaining of different kinds of expertise by developing attachments—shared commitments to the issues of housing justice and dependencies on voluntary labor. In developing these attachments, we looked to focus on the ways in which different computing systems and artifacts were part of the effort to create and articulate affective connections within the city's larger activist network. Social media, text messaging, and direct phone calls were each deployed strategically to widen the network and very explicitly to augment the group's capacity to act collectively within tightly constrained moments of protest and eviction emergency response (Asad and Le Dantec 2015). These uses of computing, which developed and transformed through the design workshops and interventions, evolved the public so that its origins in the specific and acute issues of housing justice were linked via shared attachments into a larger network of activists focused on issues of social justice in a national movement organized around issues of race, poverty, and policing that took hold following the shooting death of Michael Brown by police in Ferguson, Missouri, in early 2015.

As a public, the activists aligned with a shared set of issues, issues that were mutable and shifting as social conditions shifted. The origin in the housing crisis resulted in certain kinds of actions and the development of certain kinds of capacities to act. As that crisis abated—or more accurately, was subsumed into the widespread visceral response to newly visible acts of systemic physical violence against poor communities of color—the public likewise evolved in response to a different set of priorities among the many issues of social justice that might be pursued and contested.

* * *

Beyond these projects that continue to press on the formation of issues, articulation of attachments, and work of infrastructuring, the field of human-computer interaction plays host to a diverse collection of projects that could be productively framed or reframed through the lens of publics. Such a framing highlights different points of intervention and action via issues, attachments, and infrastructuring than the traditional footholds of productivity and efficiency or even the more recent engagement with embodiment and ludic design. Research like Lilly Irani's work examining crowd labor practices, and the ways computing might be deployed to empower workers on the Mechanical Turk platform, point out issues of labor equity and the attachments that make visible the conditions of crowdsourced labor (Irani 2015a; Irani 2015b; Irani and Silberman 2013). Silvia Lindtner's work rethinking modes of production and the culturally specific ways that making organizes local forms of invention in a global context points at ways in which infrastructuring to support local and small-scale production might further diffuse the economic and material impact of computing (for recent examples see Lindtner 2013; Lindtner, Hertz, and Dourish 2014; Lindtner and Li 2012). The research of Andrea Parker (née Grimes) examining the link between food access and healthful eating activated a network of attachments by celebrating cultural and social relationships to food (Grimes et al. 2008; Parker et al. 2012; Grimes and Harper 2008).

Parker's work in particular, when viewed through the lens of publics (with issues, attachments, and infrastructuring), casts a different light on issues of health and wellness through food as a positive experience (rather than one in need of correction or behavior modification). Issues of food access, food quality, and diet became avenues to develop attachments within the community; her work developed a research and design program that invited reflection on and celebration of food. By recasting the issue of healthful eating as one that could tap into the celebratory aspects of food—through preparation, social gatherings, and individual and collective histories with particular dishes and rituals—Parker opened a design space in which positive attachments could form in place of the corrective or coercive patterns associated with "fixing" a community. By working with the community to design a series of public interventions, her work built attachments and engaged in infrastructuring in order to share, develop, and support local action and knowledge about healthful eating, connecting these issues to the practices already present in the community.

Projects with civic outcomes in mind naturally fit under the organizing principles of publics. Whether focused on supporting deliberative action (e.g., Kriplean et al. 2007; Borning et al. 2005), or working through the creation of new technologies to support civic engagement (e.g., Gordon and Koo 2008; Baldwin-Philippi et al. 2014), new opportunities for design intervention open up when approached through issues, attachments, and infrastructuring. Recent work out of the Engagement Lab at Emerson College can be constructively considered as engaged in constituting publics (Baldwin-Philippi et al. 2014). In drawing on action research and producing a systematic approach to engaging communities in the creation of new forms of *civic media*—a term du jour for the integration of computing and other communication media for purposes of civic engagement—it shares a common intellectual heritage with the participatory design practices I've examined in this book. Both seek to cultivate solutions from *within* communities, in which issues form the basis for collective action, attachments bond a network of social and technical capacities, and infrastructuring provides a forward-looking set of capacities that can continue to act into the future.

Publics, Communities, and Commons

These different cases serve to illustrate my larger argument that publics provide both an analytic lens for understanding how design operates within community settings and specific points of entry into a community-sited design program in which achieving social outcomes through social design is the motivation. That each of these cases of social design—homeless care provision, data-driven transportation planning, community heritage, social activism—can be fruitfully considered through the lens of publics reflects in part the present moment of computing research, as it continues to explore and intervene in the world beyond the set pieces of home or work or leisure.

As I use publics to direct attention to design in community settings—specifically participatory design practice and a social design agenda—it is worth returning to the connections between community, public, and what other scholars have begun to label commons (Gilbert 2013). The communities around which my analysis has been constructed are what Carroll (2001) refers to as moral communities, in which shared obligations and common purpose bind people together—these obligations are the cultivated attachments and responses to shared issues. As design interventions are deployed in these settings, through the capacity building that comes by way of infrastructuring, the communities begin to develop the normative shared meanings of practice (Carroll 2001; McMillan and Chavis 1996). The difference between a public and a community, however, is that a public retains a mutability in identity and action: while "the danger of the idea of 'community' is that [it] too often implies a form of collectivity which is dependent upon a shared, but static and homogenous identity" (Gilbert 2013, 164), a public assumes heterogeneity and fluidity in its composition and in the ways it acts in the world. This mutability is directly connected to the privileging of issues over stakeholders—who in the same way are characterized by a stable set of relations and perspectives (Marres 2007).

So while a public operates within, or parallel to, different kinds of communities, the way it forms through and mobilizes design is orthogonal to the identity politics typically bound up in what we might normally think of as community—proximate or otherwise. McCarthy and Wright (2015, 103), provide a nice way of evincing the way community operates in and through design: that is, through a dialogical process that recognizes the value of those being engaged. This dialog creates the affective ties that enable homeless mothers and care providers, or concerned cyclists and government employees, or neighborhood residents and the neighborhood's diaspora, or social justice activists with diverse allegiances, to come together collectively and engage with a common set of issues, to cultivate a shared set of commitments and dependencies, and to build capacities for future action to earnestly address the problem originally faced by Dewey's public: that the "machine age" moves too quickly for democratic institutions to keep pace with modern issues and that the populace is no longer adequately represented or empowered to act.

Confronting—through design—the contemporary social, political, and economic issues requires new ways of constituting collective action, ways that actively seek to engage participation through multiple subjectivities and not simply through the standard computing position of *user*. Just as McCarthy and Wright (ibid.) argue that participation in co-design projects is not about turning everyone into a designer, but about incorporating and empowering multiple subjectivities to participate equally in a project of design, so too do we need to assert that computing encounters the world through a plurality of subjectivities of which user may be the least important.

The multiple subjectivities and forms of participation that circulate within a public can be distilled into alternative constructions of coproduction that "[designate] an approach to public-service management which recognizes that services are not merely 'delivered' by 'producers' to 'user' (or 'customers') ... [but] instead that service outcomes are the product of collaborative and creative relationships between professionals and members of the wider public" (Gilbert 2013, 168–169). Gilbert's larger point is that when we reconfigure social and public services around coproduction—or when we organize design around the fluid methods of co- and participatory design—we are enabling invention through a creative process that looks at social relations as generative, rather than simply as representing service targets to be met (ibid., 168). It is, however, important to further point out that these configurations are not simply about shifting responsibility from institution to individual, but about actively seeking new ways to enmesh the two around larger issues and shared attachments so that the work of infrastructuring can be undertaken.

To take a more radical position on the role of design in constituting publics, and turning one last time to Gilbert, "if the twin imperatives of neoliberalism are the reduction of all relationships to those of the market place and the reduction of all social situations to a condition of disaggregated individualisation, then the production of 'social forums'—or of 'occupations' or 'public assemblies' which share all of their characteristics—can be understood as the most basic and necessary form of resistance to this process" (ibid., 177). My suggestion here is that the conscientious creation of publics—designing publics—can become a site for countering the prevailing trends turning all aspects of our lives into market relations in which the subjective experiences of society are reduced to the roles of either producer or consumer. Those publics then design—as designing publics—and provide a pragmatic way to contest these subject positions. It is an agenda of social design through which social services, or forms of advocacy, or concerns of heritage preservation, or protest and civil disobedience, become sites for collective intervention.

Notes

Chapter 1

1. The suggestion here is not that the workplace is static or stable, nor that community settings are uniquely dynamic, but simply that there are important differences between these two settings in how people organize themselves and are organized by others.

Chapter 3

1. To be clear, relying on personal connections is not a handicap per se. The same occurs in the medical profession as doctors refer patients within a professional network based on personal knowledge of the other doctor. The parallels between health care services and social services (which often include health care) are very strong; the difference, however, comes down to higher rates of employee turnover in the social services (Le Dantec and Edwards 2008b; Voida, Harmon, and Al-Ani 2011), which frustrates the building of stable relationship between service providers and the population they serve.

2. Many of the individuals I worked with over the four years were dealing with diabetes and its effects, and with how the irregularity and instability of eviction, living with friends and family, and finally landing in a shelter confounded what is an already onerous disease to manage.

3. The HMIS being one such system, in which how the system represented the homeless individuals provided or disqualified them from access to different services. As other services moved online, similar effects were rippling out in everything from food aid to unemployment to school registration. Such systems need not be detrimental to social service provision, but for a vulnerable and disempowered population they are opaque and unassailable, and small errors that are bound to occur have substantial consequences with few opportunities for correction and remediation.

Chapter 4

1. This difference was made clear to me in a conversation with Melissa Gregg, coeditor of *The Affect Theory Reader* (2010). The rising bile of being yelled at—or being threatened with being yelled at—for no reason followed by the articulated emotions of my response to the experience traces the transition from affective response to emotional response.

2. It is worth noting that as part of the project I had a substantial number of mobile phones to provide to shelter residents, but the vast majority of the women I encountered at the shelter had their own mobile phones.

3. Passing by these issues so quickly does disservice to the significant role they play in family homelessness. Poverty was unquestionably the cause of homelessness among the women at this shelter, and no several-week program would solve their poverty.

4. A bit more background here that has been documented elsewhere (e.g., Le Dantec et al. 2010; Le Dantec et al. 2011): the shelter was closed from 8 a.m. to 5 p.m. every day, so the mothers and their children would need to leave, usually to go to work and school, and not return until the early evening. Direct contact time with shelter staff regarding programs and services was in the evenings, after dinner, while the children were working with tutors or other volunteers at the shelter. This structure meant that direct one-on-one time between staff and residents at the shelter was highly limited and already structured around the various counseling and skills-development sessions run at the shelter.

5. This despite the fact that all elements of the system were co-designed by the collection of staff and residents, so that decisions about visibility of information among staff were vetted and agreed to by residents, and categories of content for the Shared Message Board were co-developed based on the staff and residents co-constructing issues and topics that were relevant internally and externally to shelter life.

6. It should be noted that residents' messages posted to the board were not attributed to any specific resident; the lack of attribution was designed to preserve a measure of anonymity. However, within the small community at the shelter, the anonymity was not terribly strong.

7. Sadly, a brief reprieve was the most frequent outcome, as substantially redirecting what is often multigenerational poverty is not the work of a few weeks' residence at an emergency shelter. Despite this, and the periodic return of some residents, the staff were committed to treating each instance as a chance to constructively intervene in the mothers' lives, with the aim that small changes from without and from within would accumulate to alter the trajectory of the women's lives with whom they worked.

8. I have previously written about the details of how this transpired (Le Dantec et al. 2010; Le Dantec et al. 2011; Le Dantec 2012). The larger point here is that even though messages were sent directly to their phones, the move to text messaging enabled the mothers some choice as to when and how to respond to requests and direction from the staff. This newfound freedom enabled forms of social negotiation that had not previously been present at the shelter and returned a measure of personal agency to the mothers in the shelter's care.

Chapter 5

1. "Practice of community" is intentionally turned around to disambiguate from Lave and Wenger's notion of "community of practice" (Wenger 1998). The main difference is that I am more concerned with the practices themselves than with the practitioners.

2. Privileged access to the residents included access governed by privacy laws and requirements of specific social service programs, but also included the kind of social privilege of gaining the trust and acceptance of the residents. As one of the caseworkers pointed out during early fieldwork, the biggest challenge she faced was gaining the trust of the residents; once that happened, coaching them through different programs was "easy."

3. This is not to suggest that the staff were not cultivating close relationships with the residents; however, the kinds of outcomes they worked to achieve with the technology did not require strong interpersonal bonds. The mothers who came to the shelter all needed help in finding information about the right services and navigating those services to reestablish stability, and they needed this help regardless of whether they allowed staff to become close confidants.

4. "Design games" is a term coined by Ehn relating the Wittgensteinian notion of language games to design (Ehn 1988; Wittgenstein 2009). Put very briefly, language games for Wittgenstein meant focusing on words in use, i.e., the practice of language, in which the tacit rules and social and cultural setting enabled communication and signals of membership beyond the meaning of the words. Design games are a way of relating different language games, bridging between the practices present in the design setting (typically a professional setting in early work) and the practices of design. The relevance of design games here is that the connection is not simply about designers learning the language games of their users (as might be proposed in formulations of user-centered design), but about a two-way exchange in which design becomes a hybrid practice across both constituencies.

References

Alexander, S. J., Phil Edwards, Karen Fisher, and Julie Hersberger. 2005. Homelessness in Eastern King County: Information Flow, Human Service Needs, and Pivotal Interventions. University of Washington.

Alta Planning + Design. 2013. Cycle Atlanta: Phase 1.0 Study.

Asad, Mariam, Sarah Fox, and Christopher A. Le Dantec. 2014. Speculative Activist Technologies. In *iConference 2014 Proceedings: Breaking Down Walls. Culture— Context—Computing*, ed. Maxi Kindling and Elke Greifeneder.

Asad, Mariam, and Christopher A. Le Dantec. 2015. Illegitimate Civic Participation: Supporting Community Activists on the Ground. In *CSCW '15: Proceedings of the 18th ACM Conference on Computer Supported Cooperative Work & Social Computing*, 1694–1703. New York: ACM.

Asen, Robert. 2003. The Multiple Mr. Dewey: Multiple Publics and Permeable Borders in John Dewey's Theory of the Public Sphere. *Argumentation and Advocacy* 39 (3): 174–188.

Augé, Marc. 1995. *Non-Places: Introduction to an Anthropology of Supermodernity*. London: Verso Books.

Baldwin-Philippi, Jessica, Eric Gordon, Nigel Jacob, and Chris Osgood. 2014. Design Action Research with Government.

Balka, Ellen. 2006. Inside the Belly of the Beast: the Challenges and Successes of a Reformist Participatory Agenda. In *PDC '06: Proceedings of the Ninth Conference on Participatory Design*, 134–143. New York: ACM.

Bardram, Jakob E. 1998. Designing for the Dynamics of Cooperative Work Activities. In *CSCW '98: Proceedings of the 1998 ACM Conference on Computer Supported Cooperative Work*, 89–98.

Bayea, Wayne, Christine Geith, and Charles McKeown. 2009. Place Making through Participatory Planning. In *Handbook of Research on Urban Informatics*, ed. Marcus Foth, 55–67. New York: Information Science Reference.

Beck, Eevi E. 2002. P for Political: Participation Is Not Enough. *Scandinavian Journal of Information Systems* 14 (1): 77–92.

Beegle, Donna M. 2003. Overcoming the Silence of Generational Poverty. *Talking Points* 15 (1): 11–20.

Bennett, Lance, and Alexandra Segerberg. 2012. The Logic of Connective Action. *Information, Communication and Society* 15 (5): 739–768.

Bijker, Wiebe E. 1995. *Of Bicycles, Bakelites and Bulbs: Toward a Theory of Sociotechnical Change*. Cambridge, MA: MIT Press.

Binder, Thomas, Giorgio de Michelis, Pelle Ehn, Giulio Jacucci, Per Linde, and Ina Wagner. 2011. *Design Things*. Cambridge, MA: MIT Press.

Björgvinsson, Erling, Pelle Ehn, and Per-Anders Hillgren. 2010. Participatory Design and "Democratizing Innovation." In *PDC '10: Proceedings of the 11th Biennial Participatory Design Conference*, 41–50. New York: ACM.

Björgvinsson, Erling, Pelle Ehn, and Per-Anders Hillgren. 2012. Design Things and Design Thinking: Contemporary Participatory Design Challenges. *Design Issues* 28 (3): 101–116.

Borning, Alan, Batya Friedman, Janet L. Davis, and Peyina Lin. 2005. Informing Public Deliberation: Value Sensitive Design of Indicators for a Large-Scale Urban Simulation. *Proceedings of the 9th European Conference on Computer-Supported Cooperative Work*, 449–468.

Bruckman, Amy S. 2006. A New Perspective on "Community" and Its Implications for Computer-Mediated Communication Systems. In *CHI '06: Extended Abstracts on Human Factors in Computing Systems*, 616–621. Montreal: ACM.

Buchanan, Richard. 2001. Design Research and the New Learning. *Design Issues* 17 (4): 3–23.

Burrell, Jenna. 2012. *Invisible Users: Youth in the Internet Cafés of Urban Ghana*. Cambridge, MA: MIT Press.

Calhoun, Craig, ed. 1993. *Habermas and the Public Sphere*. Cambridge, MA: MIT Press.

Callon, Michel. 1980. The State and Technical Innovation: A Case Study of the Electrical Vehicle in France. *Research Policy* 9 (4): 358–376.

Callon, Michel. 2004. The Role of Hybrid Communities and Socio-Technical Arrangements in the Participatory Design. *Journal of the Center for Information Studies* 5 (3): 3–10.

Callon, Michel, Pierre Lascoumes, and Yannick Barthe. 2009. *Acting in an Uncertain World: An Essay on Technical Democracy*. Cambridge, MA: MIT Press.

Carroll, John M. 2001. Community Computing as Human-Computer Interaction. *Behaviour and Information Technology* 20 (5): 307–314.

Carroll, John M. 2004. Completing Design in Use: Closing the Appropriation Cycle. In *Proceedings of the 12th European Conference on Information Systems (ECIS 2004).* Turku, Finland.

Carroll, John M., and Craig H. Ganoe. 2009. Supporting Community with Location-Sensitive Mobile Applications. In *Handbook of Research on Urban Informatics,* ed. Marcus Foth, 339–552. New York: Information Science Reference.

Carroll, John M., and Mary B. Rosson. 2003. A Trajectory for Community Networks. *Information Society* 19 (5): 381–393.

Carroll, John M., and Mary B. Rosson. 2007. Participatory Design in Community Informatics. *Design Studies* 28 (3): 244–261.

Chick, Anne, and Paul Micklethwaite. 2011. *Design for Sustainable Change.* Lausanne, Switzerland: AVA Publishing.

Clarke, Adele E. 2005. *Situational Analysis: Grounded Theory after the Postmodern Turn.* Thousand Oaks, CA: Sage Publications.

Clough, Patricia Ticineto, and Jean Halley. 2007. *The Affective Turn: Theorizing the Social.* Durham, NC: Duke University Press.

Cohen, Kris R. 2005. Who We Talk about When We Talk about Users. In *EPIC '05: Ethnographic Praxis in Industry Conference Proceedings,* 9–30.

Conley, Dalton Clark. 1996. Getting It Together: Social and Institutional Obstacles to Getting Off the Streets. *Sociological Forum* 11 (1): 25–40.

Cross, Nigel. 2001. Designerly Ways of Knowing: Design Discipline versus Design Science. *Design Issues* 17 (3): 49–53.

De Cindio, Fiorella, Ines Di Loreto, and Cristian Peraboni. 2009. Moments and Modes for Triggering Civic Participation at the Urban Level. In *Handbook of Research on Urban Informatics,* ed. Marcus Foth, 97–113. New York: Information Science Reference.

Deleuze, Gilles, and Félix Guattari. 1987. *A Thousand Plateaus: Capitalism and Schizophrenia.* Trans. Brian Massumi. Minneapolis: University of Minnesota Press.

Dewey, John. 1954 [1927]. *The Public and Its Problems.* New York: H. Holt.

DiSalvo, Carl Francis. 2009. Design and the Construction of Publics. *Design Issues* 25 (1): 48–63.

DiSalvo, Carl Francis. 2012. *Adversarial Design.* Cambridge, MA: MIT Press.

DiSalvo, Carl Francis, and Jonathan Lukens. 2009. Towards a Critical Technological Fluency: the Confluence of Speculative Design and Community Technology Programs. In *DAC '09: DAC Proceedings of the Digital and Arts and Culture Conference*, 1–5.

DiSalvo, Carl Francis, and Janet Vertesi. 2007. Imaging the City: Exploring the Practices and Technologies of Representing the Urban Environment in HCI. In *CHI '07: Extended Abstracts on Human Factors in Computing Systems*, 2829–2932.

Doran, Derek, Swapna Gokhale, and Aldo Dagnino. 2013. Human Sensing for Smart Cities. In *Proceedings of the 2013 IEEE/ACM International Conference on Advances in Social Networks Analysis and Mining*, 1323–1330. New York: ACM.

Dorst, Kees. 2006. Design Problems and Design Paradoxes. *Design Issues* 22 (3): 4–17.

Dorst, Kees. 2015. *Frame Innovation: Create New Thinking by Design*. Cambridge, MA: MIT Press.

Dorst, Kees, and Nigel Cross. 2001. Creativity in the Design Process: Co-Evolution of Problem-Solution. *Design Studies* 55 (5): 425–437.

Dryzek, John S. 2009. Promethean Elites Encounter Precautionary Publics: The Case of GM Foods. *Science, Technology and Human Values* 34 (3): 263–288.

Dunne, Anthony, and Fiona Raby. 2013. *Speculative Everything: Design, Fiction, and Social Dreaming*. Cambridge, MA: MIT Press.

Ehn, Pelle. 1988. *Work-Oriented Design of Computer Artifacts*. 2nd ed. Stockholm: Arbetslivscentrum.

Ehn, Pelle. 2008. Participation in Design Things. In *Proceedings of the Tenth Anniversary Conference on Participatory Design 2008*, 92–101. Indianapolis: Indiana University.

Ehn, Pelle, and Morten Kyng. 1991. Cardboard Computers: Mocking-It-Up or Hand-on the Future. In *Design at Work: Cooperative Design of Computer Systems*, ed. Joan Greenbaum and Morten Kyng, 169–195. Hillsdale, NJ: L. Erlbaum.

Ehn, Pelle, Elisabet M. Nilsson, and Richard Topgaard, eds. 2014. *Making Futures: Marginal Notes on Innovation, Design, and Democracy*. Cambridge, MA: MIT Press.

Erickson, Thomas, Mark Podlaseck, Sambit Sahu, Jing D. Dai, Tian Chao, and Milind Naphade. 2012. The Dubuque Water Portal: Evaluation of the Uptake, Use and Impact of Residential Water Consumption Feedback. In *Proceedings of the SIGCHI Conference on Human Factors in Computing Systems*, 675–684. New York: ACM.

Fischer, Gerhard, and Elisa Giaccardi. 2006. Meta-Design: A Framework for the Future of End-User Development. In *Human-Computer Interaction Series*, ed. Henry

Lieberman, Fabio Paternò, and Volker Wulf, 9:427–457. Dordrecht: Springer Netherlands.

Fischer, Gerhard, and Eric Scharff. 2000. Meta-Design: Design for Designers. In *DIS '00: Proceedings of the 3rd Conference on Designing Interactive Systems: Processes, Practices, Methods, and Techniques*, 396–405. New York: ACM.

Fox, Sarah, and Christopher A. Le Dantec. 2014. Community Historians: Scaffolding Community Engagement through Culture and Heritage. In *DIS '14: Proceedings of the 2014 Conference on Designing Interactive Systems*, 785–794. New York: ACM.

Fraser, Nancy. 1993. Rethinking the Publics Sphere: A Contribution to the Critique of Actually Existing Democracy. In *Habermas and the Public Sphere*, ed. Craig Calhoun, 109–142. Cambridge, MA: MIT Press.

Friedman, Batya, and Peter H. Kahn Jr. 2008. Laying the Foundations for Public Participation and Value Advocacy: Interaction Design for a Large Scale Urban Simulation. In *The Proceedings of the 9th Annual International Digital Government Research Conference*, 305–314.

Fuad-Luke, Alastair. 2009. *Design Activism: Beautiful Strangeness for a Sustainable World*. London: Routledge.

Fung, Archon. 2006. *Empowered Participation: Reinventing Urban Democracy*. Princeton, NJ: Princeton University Press.

Gaver, William W., Jacob Beaver, and Steve Benford. 2003. Ambiguity as a Resource for Design. In *CHI '03: Proceedings of the SIGCHI Conference on Human Factors in Computing Systems*, 233. New York: ACM.

Gilbert, Jeremy. 2013. *Common Ground: Democracy and Collectivity in an Age of Individualism*. London: Pluto Press.

Gomart, Emilie, and Antoine Hennion. 1999. A Sociology of Attachment: Music Amateurs, Drug Users. In *Actor Network Theory and After*, ed. John Law and John Hassard, 220–247. Oxford, UK: Blackwell Publishing.

Gordon, Eric, and Gene Koo. 2008. Placeworlds: Using Virtual Worlds to Foster Civic Engagement. *Space and Culture* 11 (3): 204–221.

Gregg, Melissa, and Gregory J. Seigworth, eds. 2010. *The Affect Theory Reader*. Durham, NC: Duke University Press.

Gregg, Melissa. 2011. *Work's Intimacy*. Cambridge, UK: Polity Press.

Grimes, Andrea, Martin Bednar, Jay David Bolter, and Rebecca E. Grinter. 2008. EatWell: Sharing Nutrition-Related Memories in a Low-Income Community. In *CSCW '08: Proceedings of the ACM 2008 Conference on Computer Supported Cooperative Work*, 87–96. San Diego.

Grimes, Andrea, and Richard Harper. 2008. Celebratory Technology: New Directions for Food Research in Hci. In *CHI '08: Proceeding of the Twenty-Sixth Annual SIGCHI Conference on Human Factors in Computing Systems*, 467–476. New York: ACM.

Grint, Keith, and Steven K. Worden. 1992. Computers, Guns, and Roses: What's Social about Being Shot? *Science, Technology, and Human Values* 17 (3): 366–380.

Grinter, Rebecca E., and M. Eldridge. 2001. Y Do Tngrs Luv 2 Txt Msg? In *Proceedings of the 7th European Conference on Computer-Supported Cooperative Work (ECSCW)*, 219.

Grossberg, Lawrence. 2010. Affect's Future: Rediscovering the Virtual in the Actual. In *The Affect Theory Reader*, ed. Melissa Gregg and Gregory J. Seigworth, 309–338. Durham, NC: Duke University Press.

Habermas, Jürgen. 1991. *The Structural Transformation of the Public Sphere: An Inquiry Into a Category of Bourgeois Society*. Cambridge, MA: MIT Press.

Hacker, Kenneth L., and Jan van Dijk, eds. 2001. *Digital Democracy: Issues of Theory and Practice*. Thousand Oaks, CA: Sage Publications.

Hampton, Keith. 2010. Grieving for a Lost Network: Collective Action in a Wired Suburb. *Information Society* 19 (5): 417–428.

Hampton, Keith, and Barry Wellman. 2003. Neighboring in Netville: How the Internet Supports Community and Social Capital in a Wired Suburb. *City and Community* 2 (4): 277–311.

Hara, Kenya. 2007. *Designing Design*. 4th ed. Trans. M. K. Hohle and Y. Naito. Zurich: Lars Müller Publishers.

Hauser, Gerard A. 1999. *Vernacular Voices*. Columbia: University of South Carolina Press.

Hersberger, Julie. 2001. Everyday Information Needs and Information Sources of Homeless Parents. *New Review of Information Behaviour Research* 2 (November): 119–134.

Hersberger, Julie. 2002. Are the Economically Poor Information Poor? Does the Digital Divide Affect the Homeless and Access to Information? *Canadian Journal of Information and Library Science* 27 (3): 45.

Hersberger, Julie. 2003. A Qualitative Approach to Examining Information Transfer via Social Networks among Homeless Populations. *New Review of Information Behaviour Research* 4 (1): 95–108.

Hersberger, Julie. 2005. The Homeless and Information Need and Services. *Reference and User Services Quarterly* 44 (3): 199–202.

Hillgren, Per-Anders, Anna Seravalli, and Anders Emilson. 2011. Prototyping and Infrastructuring in Design for Social Innovation. *CoDesign* 7 (3–4): 169–183.

Hirsch, Tad. 2010. Water Wars: Designing a Civic Game about Water Scarcity. In *DIS '10: Proceedings of the Conference on Designing Interactive Systems*, 340–349.

Irani, Lilly. 2015a. The Cultural Work of Microwork. *New Media and Society* 17 (5): 720–739.

Irani, Lilly. 2015b. Difference and Dependence among Digital Workers: The Case of Amazon Mechanical Turk. *South Atlantic Quarterly* 114 (1): 225–234.

Irani, Lilly C., and M. Six Silberman. 2013. Turkopticon: Interrupting Worker Invisibility in Amazon Mechanical Turk. In *CHI '13: Proceedings of the SIGCHI Conference on Human Factors in Computing Systems*. New York: ACM.

Jackson, Linda A., Gretchen Barbatis, Frank Biocca, Yong Zhao, Alexander von Eye, and Hiram E. Fitzgerald. 2004. HomeNetToo: Home Internet Use in Low-Income Families: Is Access to the Internet Enough? In *Media Access: Social and Psychological Dimensions of New Technology Use*, ed. Erik P. Bucy and John E. Newhagen. Mahwah, NJ: L. Erlbaum.

Jones, Matthew R., and Wanda J. Orlikowski. 2009. Information Technology and the Dynamics of Organizational Change. In *The Oxford Handbook of Information and Communication Technologies*, ed. Chrisanthi Avgerou, Robin Mansell, Danny Quah, and Roger Silverstone, 293–313. Oxford, UK: Oxford University Press.

Klaebe, Helen, Barbara Adkins, Marcus Foth, and Greg Hearn. 2009. Embedding an Ecology Notion in the Social Production of Urban Space. In *Handbook of Research on Urban Informatics*, ed. Marcus Foth, 179–194. New York: Information Science Reference.

Kriplean, Travis L., Ivan Beschastnikh, David W. McDonald, and Scott A. Golder. 2007. Community, Consensus, Coercion, Control: Cs*W or How Policy Mediates Mass Participation. In *GROUP '07: Proceedings of the 2007 International ACM Conference on Supporting Group Work*, 167–176. New York: ACM.

Kuutti, Kari, and Liam J. Bannon. 2014. The Turn to Practice in HCI: Towards a Research Agenda. In *CHI '14: Proceedings of the 32nd Annual ACM Conference on Human Factors in Computing Systems*, 3543–3552. New York: ACM.

Larsen, Larissa, Sharon L. Harlan, Bob Bolin, Edward J. Hackett, Diane Hope, Andrew Kirby, Amy Nelson, Tom R. Rex, and Shaphard Wolf. 2012. Bonding and Bridging: Understanding the Relationship between Social Capital and Civic Action. *Journal of Planning Education and Research* 24 (1): 64–77.

Latour, Bruno. 2004. Why Has Critique Run Out of Steam? From Matters of Fact to Matters of Concern. *Critical Inquiry* 30 (2): 225–248.

Latour, Bruno. 2007. *Reassembling the Social: An Introduction to Actor-Network-Theory.* New York: Oxford University Press.

Latour, Bruno. 2008. *What Is the Style of Matters of Concern?* Assen: Van Gorcum.

Latour, Bruno, and Monique Girard Stark. 1999. Factures/Fractures: From the Concept of Network to the Concept of Attachment. *Factura* 34 (Autumn): 20–31.

Latour, Bruno, and Peter Weibel. 2005. *Making Things Public: Atmospheres of Democracy.* Cambridge, MA: MIT Press.

Lave, Jean, and Etienne Wenger. 1991. *Situated Learning: Legitimate Peripheral Participation.* New York: Cambridge University Press.

Law, John. 2004. *After Method: Mess in Social Science Research.* Annotated edition. London: Routledge.

Le Dantec, Christopher A. 2012. Participation and Publics: Supporting Community Engagement. In *CHI '12: Proceedings of the SIGCHI Conference on Human Factors in Computing Systems,* 1351–1360.

Le Dantec, Christopher A., Mariam Asad, Aditi Misra, and Kari E. Watkins. 2015. Planning with Crowdsourced Data: Rhetoric and Representation in Transportation Planning. In *CSCW '15: Proceedings of the 18th ACM Conference on Computer Supported Cooperative Work & Social Computing,* 1717–1727. New York: ACM.

Le Dantec, Christopher A., Jim E. Christensen, Mark Bailey, Robert G. Farrell, Jason B. Ellis, Catalina M. Davis, Wendy A. Kellogg, and W. Keith Edwards. 2010. A Tale of Two Publics: Democratizing Design at the Margins. In *DIS '10: Proceedings of the Conference on Designing Interactive Systems,* 11–20.

Le Dantec, Christopher A., and Carl Francis DiSalvo. 2013. Infrastructuring and the Formation of Publics in Participatory Design. *Social Studies of Science* 43 (2): 241–264.

Le Dantec, Christopher A., and W. Keith Edwards. 2008a. Designs on Dignity: Perceptions of Technology among the Homeless. In *CHI '08: Proceeding of the Twenty-Sixth Annual SIGCHI Conference on Human Factors in Computing Systems,* 627–636. New York: ACM.

Le Dantec, Christopher A., and W. Keith Edwards. 2008b. The View from the Trenches: Organization, Power, and Technology at Two Nonprofit Homeless Outreach Centers. In *CSCW '08: Proceedings of the ACM 2008 Conference on Computer Supported Cooperative Work,* 589–598. New York: ACM.

Le Dantec, Christopher A., and W. Keith Edwards. 2010. Across Boundaries of Influence and Accountability: the Multiple Scales of Public Sector Information Systems. In *CHI '10: Proceeding of the Twenty-Eighth Annual SIGCHI Conference on Human Factors in Computing Systems,* 113–122. New York: ACM.

Le Dantec, Christopher A., Robert G. Farrell, Jim E. Christensen, Mark Bailey, Jason B. Ellis, Wendy A. Kellogg, and W. Keith Edwards. 2011. Publics in Practice: Ubiquitous Computing at a Shelter for Homeless Mothers. In *CHI '11: Proceedings of the 2011 Annual Conference Extended Abstracts on Human Factors in Computing Systems*, 1687–1696. ACM.

Le Dantec, Christopher A., and Sarah Fox. 2015. Strangers at the Gate: Gaining Access, Building Rapport, and Co-Constructing Community-Based Research. In *CSCW '15: Proceedings of the 18th ACM Conference on Computer Supported Cooperative Work and Social Computing*, 1348–1358. New York: ACM.

Lindtner, Silvia. 2011. Multi-Sited Design: D.I.Y., Shanzhai and Internet Counterculture in Shanghai. In *Annual Meeting of the American Anthropological Association*.

Lindtner, Silvia. 2013. Making Subjectivities: How China's DIY Makers Remake Industrial Production, Innovation and the Self. In *Eleventh Chinese Internet Research Conference (CIRC11)*, Oxford Internet Institute, University of Oxford. Vol. 15.

Lindtner, Silvia, Judy Chen, Gillian R. Hayes, and Paul Dourish. 2011. Towards a Framework of Publics: Re-Encountering Media Sharing and Its User. *ACM Transactions on Computer-Human Interaction* 18 (5) (July):1–23.

Lindtner, Silvia, Garnet D. Hertz, and Paul Dourish. 2014. Emerging Sites of HCI Innovation: Hackerspaces, Hardware Startups and Incubators. In *CHI '14: Proceedings of the 32nd Annual ACM Conference on Human Factors in Computing Systems*, 439–448. New York: ACM.

Lindtner, Silvia, and David Li. 2012. Created in China: The Makings of China's Hackerspace Community. *Interactions* 19 (6): 18–22.

Lippmann, Walter. 1993. *The Phantom Public*. New Brunswick, NJ: Transaction Publishers.

Lipsky, Michael. 1980. *Street-Level Bureaucracy*. New York: Russell Sage Foundation.

Löwgren, Jonas, and Bo Reimer. 2013. *Collaborative Media: Production, Consumption, and Design Interventions*. Cambridge, MA: MIT Press.

Manzini, Ezio. 2015. *Design, When Everybody Designs*. Trans. Rachel Coad. Cambridge, MA: MIT Press.

Margolin, Victor, and Sylvia Margolin. 2002. A "Social Model" of Design: Issues of Practice and Research. *Design Issues* 18 (4): 24–30.

Mariano, Willoughby. 2014. Betting on "the Bluff." *Atlanta Journal Constitution*, November 1.

Marres, Noortje. 2007. The Issues Deserve More Credit: Pragmatist Contributions to the Study of Public Involvement in Controversy. *Social Studies of Science* 37 (5): 759–780.

McCarthy, John, and Peter Wright. 2004. *Technology as Experience*. Cambridge, MA: MIT Press.

McCarthy, John, and Peter Wright. 2015. *Taking [a]Part*. Cambridge, MA: MIT Press.

McMillan, David W., and David M. Chavis. 1996. Sense of Community: A Definition and Theory. *Journal of Community Psychology* 14 (1): 6–23.

Merkel, Cecelia B., Lu Xiao, Umer Farooq, Craig H. Ganoe, Roderick Lee, John M. Carroll, and Mary B. Rosson. 2004. Participatory Design in Community Computing Contexts: Tales from the Field. In *PDC 04: Proceedings of the Eighth Conference on Participatory Design*, 1:1. New York: ACM.

Mockus, Audris, Roy T. Fielding, and James D. Herbsleb. 2000. A Case Study of Open Source Development: the Apache Server. In *Proceedings of the 22nd International Conference on Software Engineering (ICSE 2000)*, 236–272.

Morelli, Nicola. 2007. Social Innovation and New Industrial Contexts: Can Designers "Industrialize" Socially Responsible Solutions? *Design Issues* 23 (4): 3–21.

Mouffe, Chantal. 2005. *The Democratic Paradox*. London: Verso.

Nelson, Harold G., and Erik Stolterman. 2002. *The Design Way: Intentional Change in an Unpredictable World*. Cambridge, MA: MIT Press.

Nicolini, Davide. 2013. *Practice Theory, Work, and Organization: An Introduction*. Oxford, UK: Oxford University Press.

Norman, Donald A. 2010. Technology First, Needs Last: The Research-Product Gulf. *Interaction* 17 (2): 38–42.

Norman, Donald A., and Roberto Verganti. 2014. Incremental and Radical Innovation: Design Research vs. Technology and Meaning Change. *Design Issues* 30 (1): 78–96.

Olivares, Teresa, Fernando Royo, and Antonio M. Ortiz. 2013. An Experimental Testbed for Smart Cities Applications. In *Proceedings of the 11th ACM International Symposium on Mobility Management and Wireless Access*, 115–118. New York: ACM.

Papanek, Victor. 1971. *Design for the Real World*. 2nd ed. Chicago: Academy Chicago Publishers.

Parker, Andrea, Vasudhara Kantroo, Hee Rin Lee, Miguel Osornio, Mansi Sharma, and Rebecca E. Grinter. 2012. Health Promotion as Activism: Building Community Capacity to Effect Social Change. In *CHI '12: Proceedings of the SIGCHI Conference on Human Factors in Computing Systems*, 99–108. ACM.

Paulos, Eric, and Elizabeth Goodman. 2004. The Familiar Stranger: Anxiety, Comfort, and Play in Public Places. In *CHI '04: Proceedings of the SIGCHI Conference on Human Factors in Computing Systems*, 223–230. New York: ACM.

Perera, Chartih, Arkady Zaslavsky, Peter Christen, and Dimitrios Georgakopoulos. 2014. Sensing as a Service Model for Smart Cities Supported by Internet of Things. *Transactions on Emerging Telecommunications Technologies* 25 (1): 81–93.

Picard, Rosalind W. 1997. *Affective Computing.* Cambridge, MA: MIT Press.

Pinkett, Randal, and Richard O'Bryant. 2003. Building Community, Empowerment and Self-Sufficiency. *Information, Communication and Society* 6 (2): 187–210.

Povinelli, Elizabeth A. 2011. *Economies of Abandonment.* Durham, NC: Duke University Press.

Putnam, Robert D. 1995. Bowling Alone: America's Declining Social Capital. *Journal of Democracy* 6 (1): 65.

Rao, Vyjayanthi Venuturupalli, Prem Krishnamurthy, and Carin Kuoni. 2014. *Speculation, Now.* Durham, NC: Duke University Press.

Ratto, Matt. 2011. Critical Making: Conceptual and Material Studies in Technology and Social Life. *Information Society* 27 (4): 252–260.

Redström, Johan. 2008. RE:Definitions of Use. *Design Studies* 29 (4): 410–423.

Reingold, Howard. 1993. *The Virtual Community: Homesteading on the Electronic Frontier.* Cambridge, MA: MIT Press.

Rice, Eric, Alex Lee, and Sean Taitt. 2011. Cell Phone Use among Homeless Youth: Potential for New Health Interventions and Research. *Journal of Urban Health: Bulletin of the New York Academy of Medicine* 88 (6): 1175–1182.

Roitman, Haggai, Jonathan Mamou, Sameep Mehta, Aharon Satt, and L. V. Subramaniam. 2012. Harnessing the Crowds for Smart City Sensing. In *Proceedings of the 1st International Workshop on Multimodal Crowd Sensing,* 17–18. New York: ACM.

Sackman, Harold. 1968. A Public Philosophy for Real Time Information Systems. In *AFIPS '68 (Fall, Part II): Proceedings of the December 9–11, 1968, Fall Joint Computer Conference,* 1491–1498. New York: ACM.

Sarpard, Kira. 2003. From Counting to Cash: How MIS Impacts the Homeless. *International Journal of Applied Management and Technology* 1 (1): 17–27.

Schmidt, Kjeld. 2014. The Concept of "Practice": What's the Point? In *COOP '14: Proceedings of the 11th International Conference on the Design of Cooperative Systems,* 427–444.

Selwyn, Neil. 2003. Apart from Technology: Understanding People's Non-Use of Information and Communication Technologies in Everyday Life. *Technology in Society* 25 (1): 99–116.

Sengers, Phoebe, Kirsten Boehner, Michael Mateas, and Geri Gay. 2008. The Disenchantment of Affect. *Personal and Ubiquitous Computing* 12 (5): 347–358.

Shapiro, Dan. 2005. Participatory Design: The Will to Succeed. In *CC '05: Proceedings of the 4th Decennial Conference on Critical Computing: Between Sense and Sensibility*, 29–38. New York: ACM.

Shirky, Clay. 2008. *Here Comes Everybody*. New York: Penguin Press.

Shirky, Clay. 2011. The Political Power of Social Media: Technology, the Public Sphere, and Political Change. *Foreign Affairs* 90 (1): 28–41.

Simon, Herbert A. 1996. *The Sciences of the Artificial*. 3rd ed. Cambridge, MA: MIT Press.

Snow, David A., and Leon Anderson. 1987. Identity Work among the Homeless: The Verbal Construction and Avowal of Personal Identities. *American Journal of Sociology* 92 (6): 1336–1371.

Spradley, James P. 1970. *You Owe Yourself a Drunk: An Ethnography of Urban Nomads*. Boston: Little, Brown.

Star, Susan Leigh, and Geoffrey Bowker. 2002. How to Infrastructure. In *The Handbook of New Media*, 151–162. London: Sage Publications.

Star, Susan Leigh, and Karen Ruhleder. 1996. Steps toward an Ecology of Infrastructure: Design and Access for Large Information Spaces. *Information Systems Research* 7 (1): 111–134.

Strauss, Anselm. 1982. Social Worlds and Legitimation Processes. *Studies in Symbolic Interaction* 4:171–190.

Strauss, Anselm. 1984. Social Worlds and Their Segmentation Process. *Studies in Symbolic Interaction* 5:123–139.

Thorsrud, E. 1970. A Strategy for Research and Social Change in Industry: A Report on the Industrial Democracy Project in Norway. *Social Sciences Information* 9 (5): 64–90.

Tuan, Yi-Fu. 1977. *Space and Place: The Perspective of Experience*. Minneapolis: University of Minnesota Press.

Turner, Fred. 2006. *From Counterculture to Cyberculture: Stewart Brand, the Whole Earth Network, and the Rise of Digital Utopianism*. Chicago: University of Chicago Press.

Vakali, Athena, Leonidas Anthopoulos, and Srdjan Krco. 2014. Smart Cities Data Streams Integration: Experimenting with Internet of Things and Social Data Flows. In *Proceedings of the 4th International Conference on Web Intelligence, Mining and Semantics (WIMS14)*, 60:1–5. New York: ACM.

Veith, Michael, Kai Schubert, and Volker Wolf. 2009. Fostering Communities in Urban Multi-Cultural Neighbourhoods: Some Methodological Reflections. In

Handbook of Research on Urban Informatics, ed. Marcus Foth, 115–130. New York: Information Science Reference.

Voida, Amy, Ellie Harmon, and Ban Al-Ani. 2011. Homebrew Databases: Complexities of Everyday Information Management in Nonprofit Organizations. In *CHI '11: Proceedings of the 2011 Annual Conference Extended Abstracts on Human Factors in Computing Systems*. New York: ACM.

Voida, Amy, Ellie Harmon, and Ban Al-Ani. 2012. Bridging between Organizations and the Public: Volunteer Coordinators' Uneasy Relationship with Social Computing. In *CHI '12: Proceedings of the SIGCHI Conference on Human Factors in Computing Systems*, 1967–1976. New York.

Warner, Michael. 2002. *Publics and Counterpublics*. Brooklyn: Zone Books.

Wenger, Etienne. 1998. *Communities of Practice: Learning, Meaning, and Identity*. New York: Cambridge University Press.

Wittgenstein, Ludwig. 2009. *Philosophical Investigations*. Ed. P. M. S. Hacker and Joachim Schulte. 4th ed. Malden, MA: Wiley-Blackwell.

Woelfer, Jill P., Amy Iverson, David G. Hendry, and Batya Friedman. 2011. Improving the Safety of Homeless Young People with Mobile Phones: Values, Form and Function. In *CHI '11: Proceedings of the 2011 Annual Conference Extended Abstracts on Human Factors in Computing Systems*, 1707–1716. New York: ACM.

Index